OLD FARMS
AN ILLUSTRATED GUIDE

Christe benedic ~ a mediaeval scribes symbol for deliverance from error.

A Welsh love spoon.

OLD FARMS

AN ILLUSTRATED GUIDE

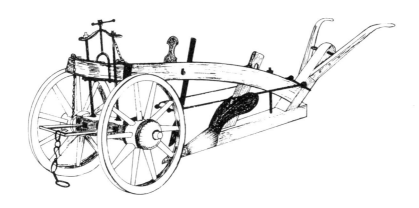

JOHN VINCE

JOHN MURRAY

© JOHN VINCE 1982

First published 1982
by John Murray (Publishers) Ltd.
50 Albemarle Street, London W1X 4BD

Printed & bound in Great Britain by the
Fakenham Press, Fakenham, Norfolk.
ISBN 0 7195 3938 1

CONTENTS

INTRODUCTION

AGRICULTURE is man's most important activity. The ancient farmers who devised techniques which produced a surplus of food laid the foundations of civilization. This surplus fed a few men who could then devote all their energies to crafts or the arts. Development of Society, science and artistic life depended in a very practical way on the farmer's skills. In his sphere, clearly defined seasons followed one another and marked out the patterns of labour. Each year brought with it the same struggles to wrestle from the land both subsistence & surplus.

The tools at the farmer's command were simple & often home made. His greatest source of power was the ox & the horse. Together they ploughed, harrowed & carted. The ox was superior at the plough, but in the C19 the horse began to rule the furrows.

For centuries the rhythm of agriculture was bound up with the manual toil of the stripholder and the patient plod of ox & horse. When the feudal system of open fields began to change and the landscape became a patchwork of quickset hedges, the old way of life took on a new aspect. Countrymen were no longer self-sufficient and many had to learn the harsh lessons of becoming day workers. In wet weather there was no work and no pay.

Before enclosures perhaps as much as one third of the arable land remained fallow each year. Poor drainage restricted grazing and the ease with which stock became infected all combined to inhibit the development of husbandry. Once the feudal bonds had been broken the enclosures made it possible for men like Turnip Townshend, Jethro Tull and later Messrs. Ransomes to effect profound changes. The former grew root crops & clover instead of letting a field lie fallow every third year. The number of cattle that could be overwintered increased. Instead of being killed, they lived off turnips when there was no grass. The second mechanised the task of sowing and made it possible to hoe the crop growing in rows with horsedrawn hoes. A self-sharpening share was developed & standard parts for the plough were made by the latter. Together these basic advances depressed the demand for field labour. Mechanical threshing replaced the flail in the 1820s/30s. The consequent disturbances and rick burning of the period have been recorded by writers like W.H. Hudson - in 'A Shepherd's Life'.

In the late C19 the mechanical reaper & the binder appeared and the demand for labour diminished further. The last two decades of Victoria's reign were a time of great hardship for the ordinary farmworker and his family. Part of their discomfort can be attributed to the effects of mechanisation, & part to the Free Trade policy which fostered cheap food imports. It has been the lot of the farmworker to be under-valued. The part he plays in creating real wealth is never acknowledged by Society until it has plunged itself into war. After the Great War of 1914-18 there was a new dearth of labour. The first combines came on the scene in the 1920s, the tractor became a regular sight, & the age-old rhythm of the fields slipped into oblivion. With the changed emphasis on mechanical methods many of the buildings also became obsolete. There was a new purpose in farming, and a new motivation ruled by a balance sheet. The growth of suburbia created a new demand for dairy products. Dairying became important as it provided a better cash flow for the farmer.

Six decades later we can appreciate anew the importance
of the countryman's home, its fittings and the tools employed in
his particular way of life. In the pages below the principal
aspects of the old way of life are illustrated. Each year
thousands of visitors go to see the many museums of rural
life that have been created, a good number by private
enterprise.

There is something elemental about re-encountering the
things you once used or saw your grandparents using.
The pace of change in all areas of Society during the past
half century has been more dramatic than anything
which went before. Through the artefacts of our ancestors we
can still touch the hem of the garment which clothed their
daily lives. When you feel the worn smoothness of the butter
churn's handle or touch the gloss on the rake's much-used
grip you are in a literal sense in contact with the past.
You cannot separate these things from those who used them
and whose hands made them warm in working.

The oldest surviving form of farmstead, which remained
in use until very recent years, is the blackhouse which had its
origins in the Viking period. A preserved example, on the
Isle of Lewis, gives us a visual contact with those remote
days. At Acton Scott, Salop, there is still a working farm
that operates on real rippling horse power. Such things
are not anachronisms, but a demonstration of man's need
to value earth's natural resources. Those with eyes to see
have much to learn from the past, and parts of it may
prove essential to sustain the future. When the oil runs
dry a plough team might still tug some sense into our
civilization.

THE FARMHOUSE & WOMEN'S WORK

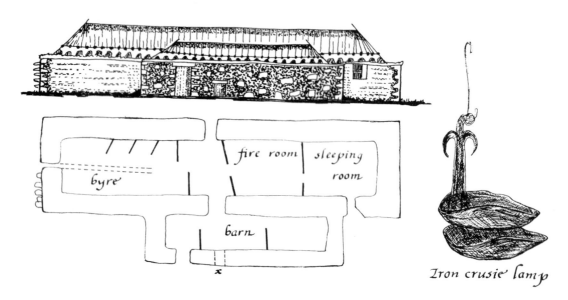

byre

fire room | sleeping room

barn

x

Iron crusie lamp

THE HEBRIDEAN BLACKHOUSE

A blackhouse provides us with a glimpse of a way of life that the Vikings knew. Man & beast shared the same roof. The longhouse tradition began in the Dark Ages. A blackhouse has a double wall with an infil of peaty soil. A simple wooden roof structure supported the layer of heathery turf which formed the inner lining of the roof. Bunches of straw were then placed on top. A net held down by heather ropes and anchor stones (acraichean) kept the straw in place. At each end of the hipped roof a stick to support the ropes can be seen. These 'raven sticks' (corra thulchainn) are reminders of the Celtic raven-goddess, a deity of fertility & war. In the Viking period raven sticks were probably often decorated with severed human heads.

Doorways placed in line provided a common passage for stock & man. A house had two parts. The byre and living area were separate as the plan shows. Family and community life focussed on the fire room. There were no chairs, although stools were used. A settle (being) stood against the wall. This left more space around the central peat fire (cagailt). An iron chain holding the pot hook or crook (dubhan) was fixed to the ridge pole. Dressers had a sloping top to suit the roof. Smoke from the fire stayed inside and soot gathered on the turf roof lining. It was once customary to remove the turf each year to provide manure for the fields.

In the outside barn wall, and in line with the doorways, there was a low arch - x - called the winnowing hole (toll fhasgnaidh).

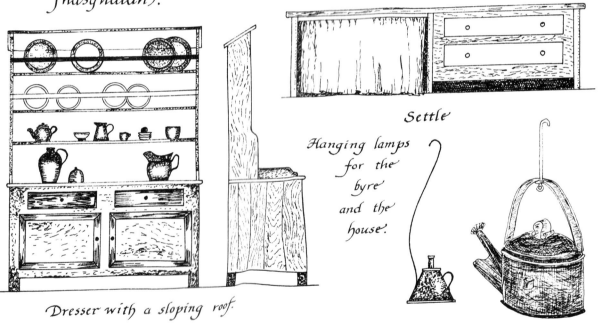

Settle

Hanging lamps for the byre and the house.

Dresser with a sloping roof.

When new building materials were introduced from the mainland in the C19 the new dwellings were called "white houses". The traditional buildings then became known as blackhouses.

THE FARM HOUSE

C18. Welsh house with turf ridge.

The tradition of the longhouse lasted for centuries. In its original form man and beast shared the same door. Although the shape remained the house part acquired its own doorway and an internal wall separated it from the byre.

Turf roof on plaited branches. Mediaeval Welsh.

A chimney of mud also added to the householder's comfort. For centuries thatch was the main roofing material and this was sometimes capped with growing turf. Longhouses were often

C14. Cruck house ~ Welsh.

built so that the byre end was at a lower level. Windows, until glass became cheaper in the C18., remained small. Simple shutters were sometimes used to cover window openings in the house. In the C18. and later years many farmhouses had larger windows added. Some small windows often survive and are worth looking for. Even blocked windows are easy to identify.

Stone slats

C17. Cumbria.

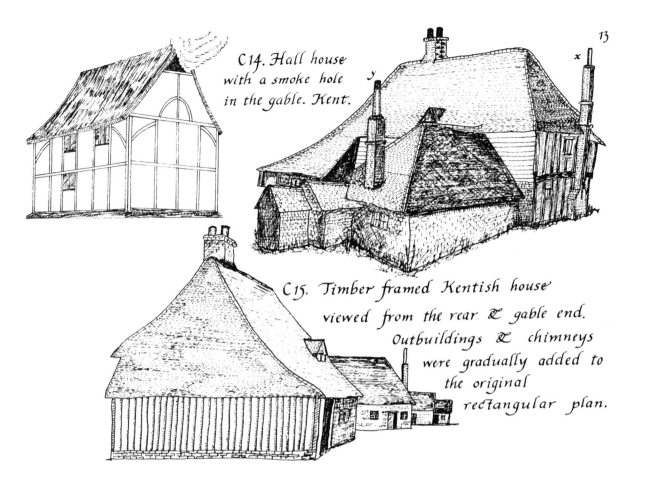

C14. Hall house with a smoke hole in the gable. Kent.

C15. Timber framed Kentish house viewed from the rear & gable end. Outbuildings & chimneys were gradually added to the original rectangular plan.

On the eastern side of England the typical farmhouse started off as a rectangular building. The simplest houses were open halls with a smoke hole. As domestic life developed many open halls were divided into separate rooms, and their mediaeval features became hidden. Reconstructed examples can be seen at museums at Avoncroft, Singleton & Stowmarket ~ see page 156. Another way of making a house larger was to extend the roof downwards to form an outshut. The C15. house shown above has had its rear elevation and one gable added to in this manner. When a timber house is externally plastered alterations are much more difficult to spot. Chimneys can be significant indicators of a building's development. The extra ones shown were for a sitting room (x) and a wash house (y).

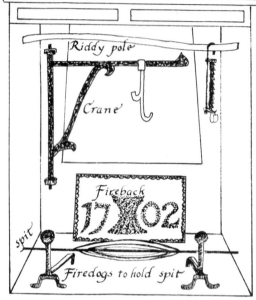

Oven brush

Mop

Faggot fork

Peel – to place or remove loaves

In the mediaeval period cooking was often kept away from the hall. As houses became divided into compartments one room was used for cooking. By the C19 larger farmhouses often had two kitchens. The lesser one, the backhouse in E. Anglia, was dedicated to menial tasks which included baking & washing. The equipment of a backhouse was very similar to that found in the mediaeval house. The focus was the fireplace, with its crane which carried the hooks, pots & kettles. In the wide chimney the riddy pole also supported a hake or chain. The adjustable hake, a term derived from the Norse "haki" meaning hook, has its origins in the C13 or before. Food was often cooked on a spit supported by two firedogs. A fireback protected the wall. Close to an old fire was a small shelf for the salt. Bread was baked in an oven set into the wall. Faggots were burnt inside to heat the brick lining. Ashes

Riddy pole

Crane

Fireback
17 02

Spit

Firedogs to hold spit

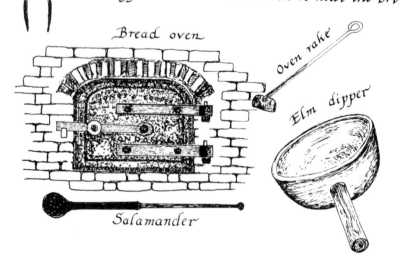

Bread oven

Oven rake

Elm dipper

Salamander

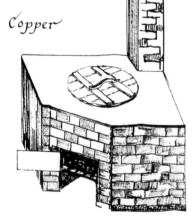

Copper

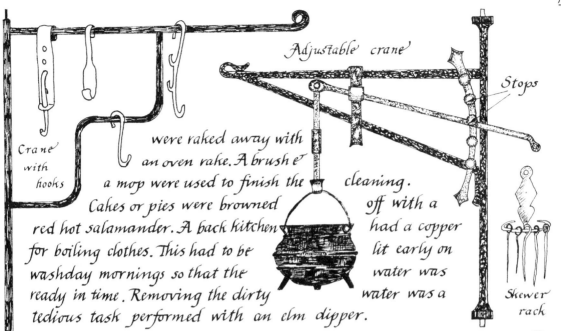

Adjustable crane

Stops

Crane with hooks

Skewer rack

were raked away with an oven rake. A brush & a mop were used to finish the cleaning. Cakes or pies were browned off with a red hot salamander. A back kitchen had a copper for boiling clothes. This had to be lit early on washday mornings so that the water was ready in time. Removing the dirty water was a tedious task performed with an elm dipper.

An idleback or kettle holder was a refinement to be found in many houses. It saved taking a heavy hot kettle from its hook to pour water. Kitchens abounded in examples of the blacksmith's skill. The most interesting objects are probably adjustable hakes or cranes. To raise or lower the pot the catch was moved so that it rested below another stop. Some utensils were always plain & functional like the griddle used for such things as oatcakes. Clay pipes were cleaned in a pipe rack among hot embers.

Hake

Old style meat skewers were made to last. They had a special skewer rack. The example shown bears a heart and diamond. Such items reflected the smith's own tastes.

Griddle

Pipe rack

WASHING & IRONING

Monday was always washday. Then the kitchen was filled with clouds of steam and the unforgettable smell of soapsuds & dripping clothes - all stirred up with a laundry bat or copper stick. Before modern detergents had been invented countrywomen had to make up their own washing aids. Woodash from the kitchen fire was strained through a lye dropper. Twigs were laid in the bottom and a piece of muslin placed on top. Water poured over the ash dripped through the holes. The resultant alkaline solution was used for flannels and whites. Rougher garments before boiling had to be scrubbed on a scrubbing board to release the dirt. The dollystick used in a dollytub was an alternative to the scrubbing board. There were several variations on the dollystick idea - these included the posser & poss stick. Before galvanised tubs came into use coopered versions were commonplace. The latter were often known as bucktubs; buck being an old dialect word for lye.

In larger farmhouses it was customary to employ two visiting washerwomen. They used to wear pattens to keep their feet clear of the wet stone floor (see p.153). After the labour of washing, water had to be squeezed from the clothes. Washing was once wound around a wooden roller and a mangle bat rolled across it. The invention of the mangle made the task of wringing out washing much easier. Stockings were dried on a stocking board to help them keep their shape.

Ironing was a slow business in the days when every iron had to be heated on a stove. The 4lb. flat iron was used for most general work. In use it could stand on a trivet while the garment was turned over. Iron holders were made at school from odds & ends. Box irons had an iron core that was made red hot and put inside. Charcoal irons were filled with charcoal. The user would swing it to and fro to keep the fuel glowing.

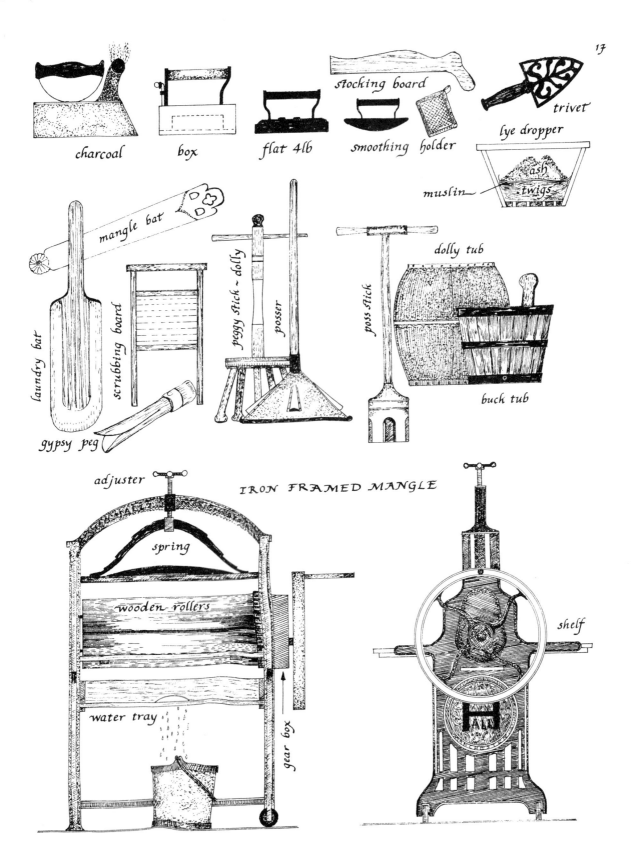

charcoal

box

flat 4lb

smoothing   holder

stocking board

trivet

lye dropper

muslin

ash twigs

mangle bat

laundry bat

scrubbing board

gypsy peg

peggy stick ~ dolly

posser

poss stick

dolly tub

buck tub

adjuster

IRON FRAMED MANGLE

spring

wooden rollers

water tray

gear box

shelf

THE GREAT KITCHEN

Towards the end of the C19 great changes came about in the nation's kitchens. Open fires gave way to cast iron kitcheners. Oven temperatures could be controlled in a way not possible on an open fire ~ the doors open towards the fire to protect it from spitting fat. There is a deep drawer on each side to keep things warm. Above the top a shelf provides additional warm storage. With equipment of this size a vast amount of cooking could be carried out under the eye of the cook who no doubt had several living-in maids to assist her. Cleaning the flues and blackleading the stove was a constant chore for hard pressed girls who once started their working lives at twelve years of age. The day in the kitchen started at five o'clock when the first task was to clear the ashes and light the fire. Most of the work took place

Windsor stool & chairs —

C17. Oak hutch

on the well scrubbed kitchen table. This was the general work-
bench and it usually held shallow drawers that could be used
for knives & other utensils. Chairs were an important part of
kitchen furniture. Windsors with elm seats were commonly to be
found in country kitchens. Their design depended upon the
date they were made. Wheelbacks were made for a long time.

Storage of food and utensils was a problem in the mediaeval
kitchen. The first cupboards were simple and belonged to the
wealthy. By Victorian times even modest households had acquired most essential items of furniture. The kitchen dresser was
very important. Plates & cups could
be kept on the shelves & the cupboard
could be used for food. Bread had
a place in an earthenware bin
which helped to keep it fresh.

Kitchen table

Flour bin   Bread bin   Dresser

Spill rack

Match box with an abrasive lid.

Strainer spoon

Trencher

Spoon rack

Grater

Salt hole

Pastry marker

Wooden platter

Earthenware strainer & lid

Potato masher

Horn spoon

Porridge stirrer

Handled bowl

Horn mugs

Glove darning egg

Sock darner

Wooden salt

Pewter salt

milking
stool

piggin

yoke

cream pan

THE DAIRY

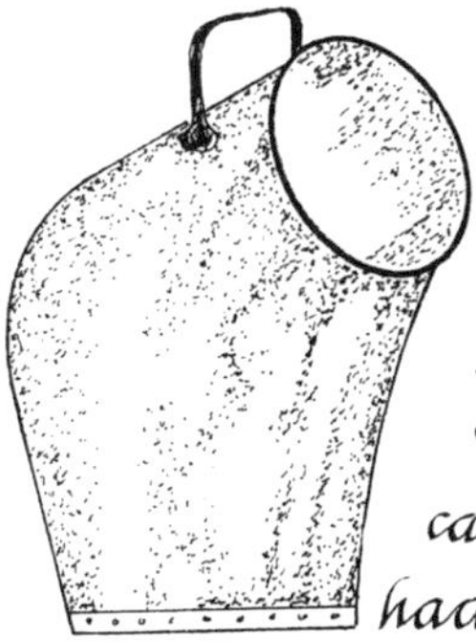

canted pail

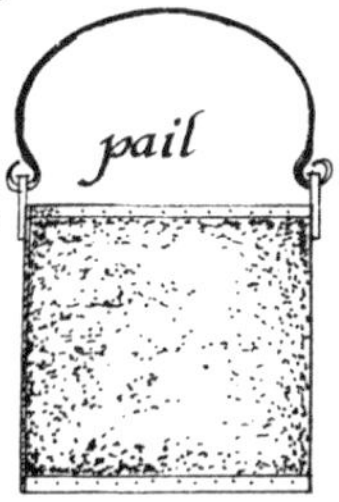

fleeter

wooden
dipper

railway
churn

pail

pint
measure

delivery
churn

Dairying played an important part in Victorian farming. As populations increased so did the demand for milk, butter & cheese. Milking by hand was an arduous task. Cows were often milked afield & the heavy buckets carried home on a yoke. Old style pails of oak had an upright handle. When sheet metal pails came into use wooden buckets were discarded. Contrary cows could easily put a foot into a bucket. The canted bucket, by Listers, was designed to reduce this risk. Measures & churns were kept scrupulously clean. There are still people who can recall when milk was delivered direct from the milkman's churn with the help of a pint measure. Even the wide-topped bottles with their card seals and perforated centres that could admit a straw are now collectors' items. Before the days of tankers milk was sent daily in churns by rail to the towns. The shape of the older style of churn is shown here. Churns of this kind can still be seen as garden ornaments ~ with geraniums where the cream used to be. Cream for butter was skimmed from the pan with a fleeter.

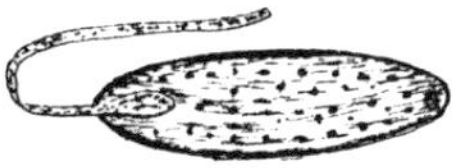

cream pot

butter cup

bottle top

There were several types of churn.
From the mediaeval period came
the plunger churn that was still
in use in the C19. By that time
stoneware models were being
made. End over end churns were in use on many
farms. The task of turning at the rate of 40 turns
a minute demanded a good deal of energy. Box
churns came in rectangular and round versions.
The cream was agitated by paddles. To stop butter sticking to
the side of the churn it was washed in a solution of brine. You
could tell when the butter had 'come' by the sound it made in the
churn. A traditional rhyme once chanted by dairymaids went
like this :~

"Come butter come, Peter stands at the gate,
Waiting for a buttered cake, Come butter come."

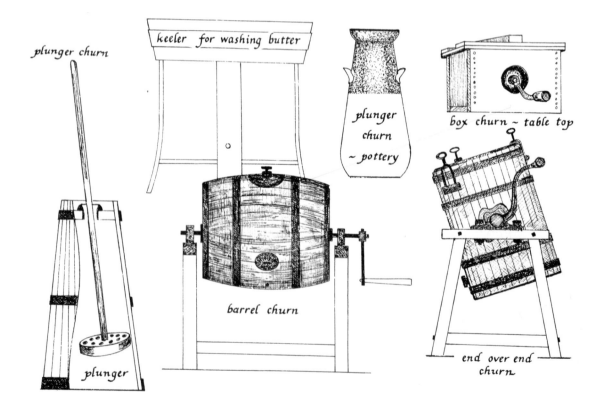

plunger churn

keeler for washing butter

plunger
churn
~ pottery

box churn ~ table top

barrel churn

plunger

end over end
churn

wooden
fleeter

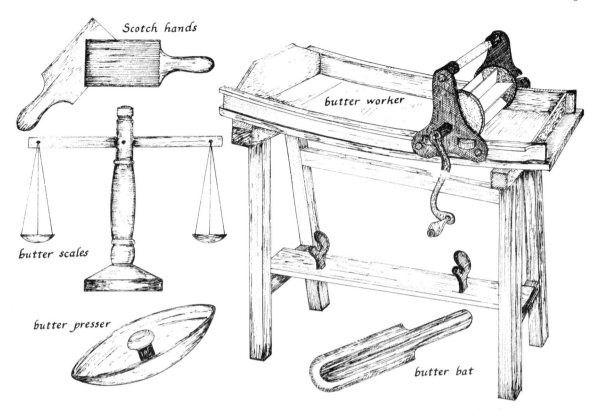

Scotch hands

butter worker

butter scales

butter presser

butter bat

Once the butter had formed inside the churn the water had to
be extracted. This was done by squeezing it on a butter board
with a presser or beater. A butter worker was invented to speed
this task. Its tray sloped to allow the whey to run off. Then the
butter was weighed and worked into blocks with Scotch hands
~ they were ribbed on one side ~ or placed in a butter mould
and decorated with a box stamp. Butter had to be made at
a precise temperature and the buttermaker's cardinal rule
was always strictly observed ~

NEVER TOUCH BUTTER WITH YOUR HANDS.

butter prints

butter mould
& print

To avoid paying Window Tax (introduced 1695) farmers could place a notice like this above the dairy door. The tax was repealed in 1851.

DAIRY.

Victorian dairy ~ Berkshire. The extended roof helped to keep the interior cool.

Milking shed ~ Wiltshire.

C 18 Dairy ~ Oxfordshire. Its Dairy notice is hidden beneath the porch.

Three piece wooden latch ~ fixed with dowels.

PART SECTION ~ CLAPSES at GROVE FARM, BIERTON, BUCKS.

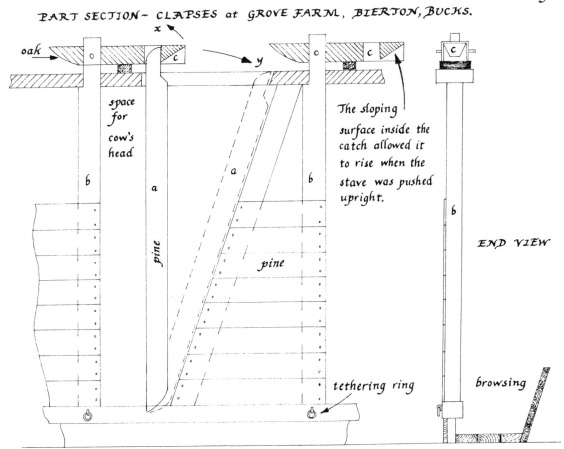

END VIEW

The sloping surface inside the catch allowed it to rise when the stave was pushed upright.

space for cow's head

tethering ring

browsing

The wooden CLAPSE is a curious but important survival. It was used to hold the cow's head during milking. A manger or browsing was placed behind the clapse. Each one had a moveable stave (a) which was held in place by an oak catch (c). When the stave was in its vertical position the distance between it and the stud (b) was exactly 7½ ins. A larger space would allow a beast to free herself. To release the stave the catch (c) was lifted in direction x which allowed the stave to fall towards y. Several clapses were placed together in a milking parlour or even a remote barn. The clapse seems to be confined to Bucks. but readers may be able to tell the author of its use in other areas. Most farmers ceased using the clapse long ago, but one set is still in daily use.

CHEESE

The names Stilton, Cheddar
& Double Gloucester remind
us of the days when most
farms made cheese. Local
tradition was important
but the same basic sequence
was followed. The milk was
coagulated by adding
rennet ~ an extract
from a calf's fourth
stomach. Green curd
was broken with curd
knives or a breaking
machine. Cutting
helped to free the whey which
was finally expelled when the
cheese, in its vat or chesset,
felt the pressure of the press.
The best presses used a system
of levers to generate a force
of 2 or 3 tons. When the
pressing was over cheese
was stored in a well ventilated
room, and each cheese was
regularly turned by the dairy
maid until it matured.

curd breaker

chesset

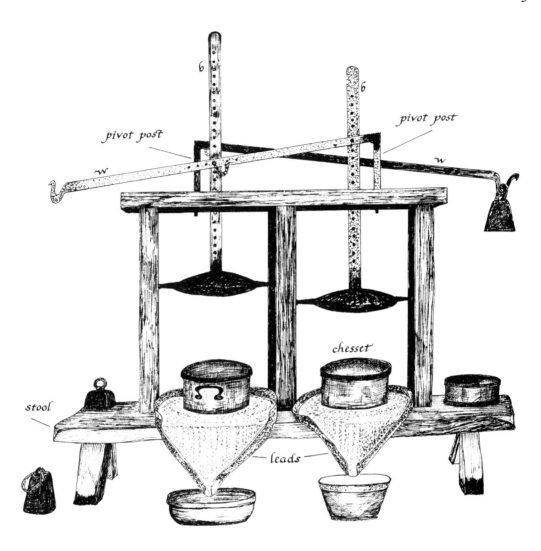

b b

pivot post pivot post

w w

chesset

stool

leads

A double cheese press with several weights. The two leads collected
the whey which oozed from the chessets and allowed it to drip into
the bowls on the floor. Wooden presses of this kind were made by
the village carpenter. Ironwork came from the blacksmith. The press
was adjusted by placing a pin through a weight lever (w) & a
press bar (b). Cheese rooms often had a northern aspect as coolness
was important where cheese was concerned.

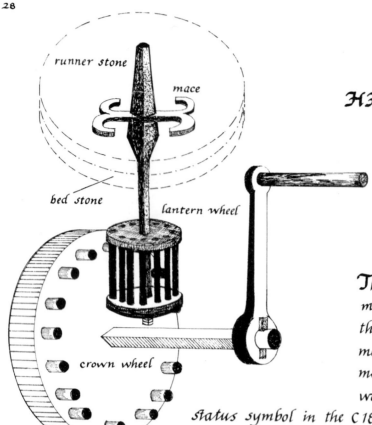

runner stone

mace

bed stone

lantern wheel

crown wheel

HANDMILLS

There were very few mechanical aids to help the housewife in the mediaeval period. The machine shown here was a considerable status symbol in the C18. Its arrangement of gears is a scaled-down version of the machinery found in a watermill. The crown-wheel is driven by a handle and as it turns the teeth engage the staves of the lantern wheel. This latter wheel gets its name from the similarity it has to a horn lantern. As the lantern wheel rotates it turns the shaft and the mace which supports the upper, runner stone. The bed stone does not move. The grain was fed into the centre, the eye, of the stone; and the flour discharged around the edge. Wooden casing around the stones, not shown in the drawing, was arranged so that the flour emerged via the chute. A staple was fixed at the front to allow a flour sack to be held.

The amount of energy needed to operate a handmill was quite considerable and its machinery needed a substantial frame to contain it. Dates and initials were sometimes carved on the framing and such details always add interest to the homespun mechanics which handmills possess. The position of the lantern wheel could be adjusted vertically to alter the distance between the stones. An adjustment was made with wedges or a screw thread. On the drawn example the wing nut raised or lowered the member (c) which is morticed into (b). In this way very exact adjustments could be made. A spare stone lies face up inside the frame so that its cutting grooves can be seen.

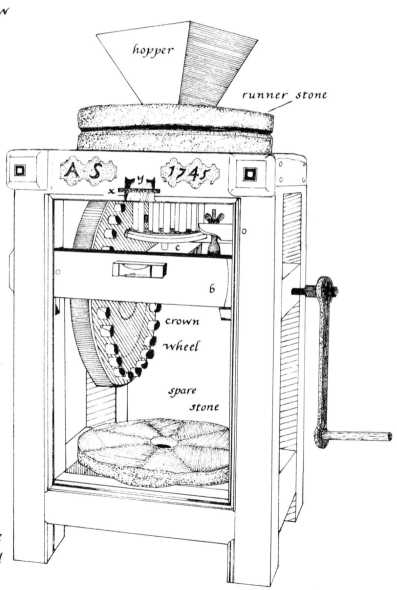

hopper

runner stone

A S 1745

crown wheel

spare stone

The grain was put into the hopper which fed it into the stones. Flour was expelled at the circumference. It then flowed through the chute (y) and into a sack held by the staple (x).

C14. Flint ~ Sussex

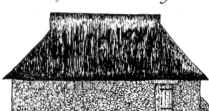

C19. Stone with render ~ Somerset

end elevation ←

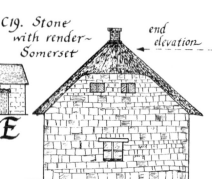

COTTAGE LIFE

We know very little about the design of cottages before the C15. In the early mediaeval period many dwellings were probably very simple structures like the turf huts used by the charcoal burners in this century. The materials used determined a building's features. In the western part of the British Isles stone houses made without mortar were fashioned in the beehive shape shown below. A pig stye of this shape can be seen at St. Fagans ~ page 156. Teapot Hall, which formerly stood in Lincolnshire, was a very basic kind of house. Its two straight crucks formed both roof and wall. This idea was improved upon by the addition of a vertical wall as the Gloucestershire example shows. At Singleton a C14. flint cottage has been reconstructed on the basis of archaeological evidence. Its thatched roof is conjectured but it must have had a smoke hole. Most of the early cottages had one room. The use of the roof space for sleeping had to wait until the chimney was added. Although bedrooms had appeared in large houses at a very early date most cottages had to wait until the C18. before they enjoyed the added comfort of an upstairs. Many cottages still bear the evidence of this development. The upper windows often rested on top of the wall.

Charcoal burners' huts.

Pig stye with a corbelled roof.

Teapot Hall

C19. Chalkstone ~ Berks.

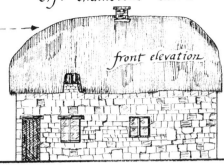

front elevation

C19. Stone ~ Oxon.

C19. extension : C18. Brick ~ Lincs.

C18. Cob ~ Devon

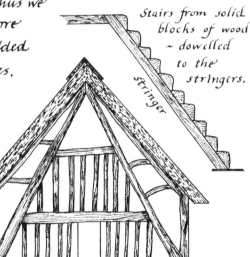

This meant that they were at floor level. The examples from Berks., Lincs., & Oxon. show windows of this kind. To reach the upstairs some steps were needed. Some very old buildings still have solid wooden steps fixed with wooden dowels. Less elaborate arrangements were made in most cottages where a simple ladder was fixed to an end wall. A trapdoor in the floor was an added refinement. Old carpenters' accounts often refer to stairholes. Eventually ladders became proper staircases and they were partitioned from the main living area. Thus we gained the 'Stairhole door'. To provide more space cottages frequently had outshuts added for the storage of fuel & other necessities.

Mud, cob, dwellings often had rounded corners for strength & a tarred base to the walls. They had deep eaves to keep the walls dry.

stair hole

ladder

Stairs from solid blocks of wood ~ dowelled to the stringers.

stringer

C15. Straight crucks & vertical walls ~ Glos.

THE COTTAGE KITCHEN

From the C18. onwards owners of large estates carried out improvements to their cottages. In some places new model villages were created. The row shown above was built at Leverton, Berks. circa 1800. Their bedroom windows are in the gable but well above the level of the front wall. Earth floors were usual in most cottages in Georgian times, and one room served all the family's needs. The fireplace was very much the same as the farmhouse kitchen, but smaller. Cooking methods were probably simpler too. Cottagers without an oven had their bread baked by the village baker. Some villages had a communal bakehouse. The fire seldom went out and a pot was always simmering or a kettle placed to warm the water.

Windsor chairs & a tripod table which sat more firmly on an uneven floor.

Gradually the cottage kitchen reached the standards set in the farmhouse. Smaller versions of the cast iron kitchener were produced and these found their way into the cottage. Very often the riddy pole and hake remained too and the mediaeval and the modern met together. Some open chimneys were too large for the smaller ranges and so they were built into one corner like the example below. The space surrounding was used of course to hang herbs to dry, and the bellows.

Many domestic items were handed down and cottagers often used cast off utensils that were no longer needed in grander homes. Objects like the skillet with its three legs that enabled it to stand in the embers of an open hearth found their way into the cottage to serve as an extra saucepan or frying pan. By removing the iron hob of the stove it could comfortably sit in the fire. Some trivets were made to fit on the firebars and so provide an extra shelf that would keep a pot simmering or a kettle hot.

Ingle nook fitted with an iron range.

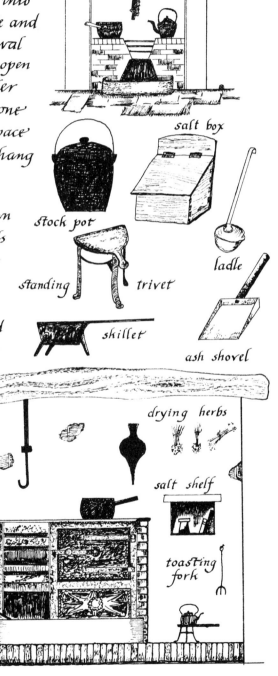

salt box

stock pot

ladle

standing trivet

skillet

ash shovel

drying herbs

salt shelf

toasting fork

PEAT

Peat was cut in many parts of the British Isles & it still forms part of the crofter's life in the Hebrides. Peat cutting tools were made to suit the user, and come in many forms. In the Islands peats is the collective term for the slabs of fuel, but in the south turf was the usual name. Cutting took place in May & the dark soggy slabs were laid out to dry. After a few drying days the peat was 'stooked' so that it was ready to be carried in the dryness of summer. In Somerset barrows were used to wheel peat to a track for carting. On the Isle of Lewis men did the carting. Women folk carried peats all the way home on their backs in wickerwork creels. Hand barrows were used on wet terrain in Wales. Three types of peat tool are shown. The instrument (a) is an iron for marking which divided the top layer of turf into squares before it was pared. The long paring iron (b) with its angled handle was used to lift the marked turf. Narrow spades with a projecting fin enabled slabs of a regular width to be cut (d, e, f, g). Arrow-headed knives (h, k) are from Cornwall & Devon. To lift dried peat a twin pronged titch crook (l) was used.

The feudal landscape was different from our own. Peat was then a basic fuel. Most people are unaware of the fact that the Norfolk Broads were formed as a result of the activities of mediaeval peat cutters. In those days TURBARY rights which allowed turf to be cut from the common land were very important. Mediaeval records frequently use the Latinised form 'turbarium' to indicate a place where turf was cut. Place names too remind us of the location of peat bogs e.g. Mosser, Moze, Chat Moss, Mosedale & Morestead. These derive from the Old English 'mos' & 'mor'. Fenham and Fencot come from O.E. 'fen' which means a boggy place. The other tools are : m - a peat knife ; n - a turf spade.

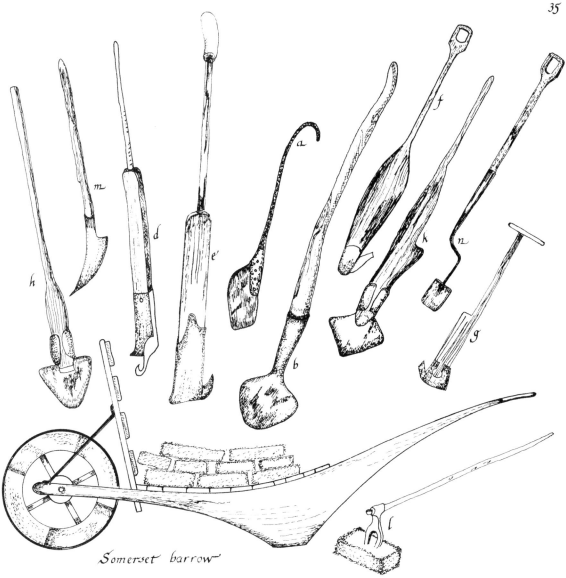

Somerset barrow

A Gaelic kindling blessing.

I will kindle my fire this morning, In presence of the holy
angels, without malice, without jealousy, without envy, without
fear of anyone under the sun.

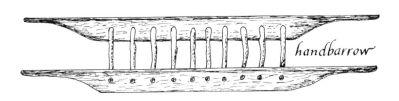

handbarrow

RVSHLIGHTS

Country people used to rise at daybreak & go to bed at dark. Lighting was expensive and used as little as possible in ordinary homes. The simplest lights were made from rushes ~Juncus effusus~ that had been peeled and their pith dipped into fat. These thin rudimentary candles were fully described by Gilbert White - see page 158. Rushlights were held in holders that had jaws shaped like pliers. Sometimes the counterweight which kept the jaws closed was formed to hold a candle. Rushlight holders were made by village blacksmiths and do not conform to a standard pattern. The unusual holder drawn here is a smaller version of the chimney crane on page 15. In use rushes burnt quite quickly.

It was a child's task to keep moving the rush in the holder so that the jaws did not put out the precious flame.

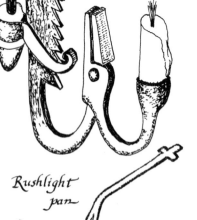

Adjustable rushlight & candle holder

Rushlight pan

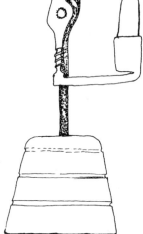

Rushlight holder

Clay candlestick

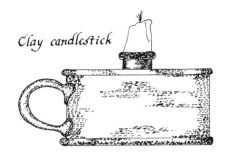

CANDLES

Horn lantern

Hundred eyes lantern

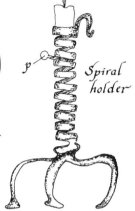

Spiral holder

Blacksmith's holders

Tin candle-stick

Wire mesh lantern

Candle Box

There is something distinctive about candle light. Although candles were costly they became more widely used after paraffin wax replaced tallow in the 1850s. An open flame was always a hazard & most candle sticks had wide bases to make them as stable as man could contrive. Blacksmiths made candlesticks for the home & for workshops. Their designs are always individual & contrast with the mass produced clay or enamel alternatives. Some holders had pushers (p) which moved the candle upwards. Lanterns made candles safer, but the open top of the 100 eyes lantern was always a risk. Horn lanterns were old even in Shakespeare's time. Cellar candlesticks were rudimentary. They had spikes or hung from a nail. The fork on the spiked holder was a 'save all' for candle ends that were later remelted.

Cellar holder

save all

Spiked holder

Tinplate holder

LONGCASE PUMPS

A water supply was a very basic need for any household. The daily life of the farmhouse required a considerable amount of water. The dairy particular used a great deal. Wells are the oldest form of man's efforts to secure a reliable supply of water, but the work of raising it a bucket at a time was very slow. We do not know the identity of the genius who developed the pump which enabled a constant stream of water to be taken from the well.

There are still many longcase pumps to be seen. Some have been restored to working order. When the handle is moved downwards the bucket in the pipe moves up; & the leather washer at its base helps to provide the sucking action which raises the water into the cistern. The bottom of the cistern is always covered by water so that the pipe does not become air locked. An iron rod connects the bucket via a shackle to the handle.

THE WORKING PARTS

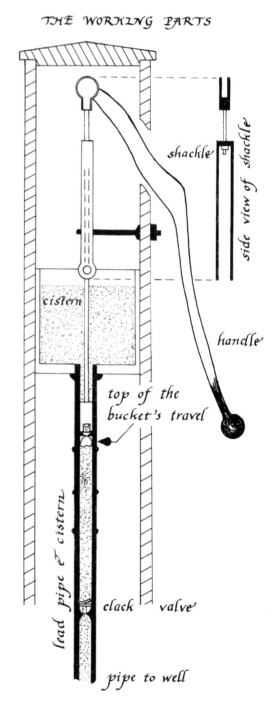

shackle

side view of shackle

handle

cistern

top of the bucket's travel

lead pipe & cistern

clack   valve

pipe to well

On very old pumps these parts were often made of wood. When the level of water in the lead cistern rises above the spout water is then discharged. To protect the mechanism the pump was boarded.

SIDE VIEW

The visible metal parts were the handle & the face of the cistern. In very cold weather pumps were protected with straw held in place with sacking and string. Even with these precautions they often had to be primed with hot water on winter mornings. Galvanised buckets were introduced in the Victorian era; but many wooden ones were used long after the cheaper versions became commonplace. Longcase pumps were not mass produced.

wooden case usually of elm or oak

spout face with date

RCE
✛ · ✛ · ✛
18 99

The oldest surviving longcase pumps seem to date from the C18. It was customary to decorate the front of the cistern. This example shows a date as well as the owner's initials.

PUMPS

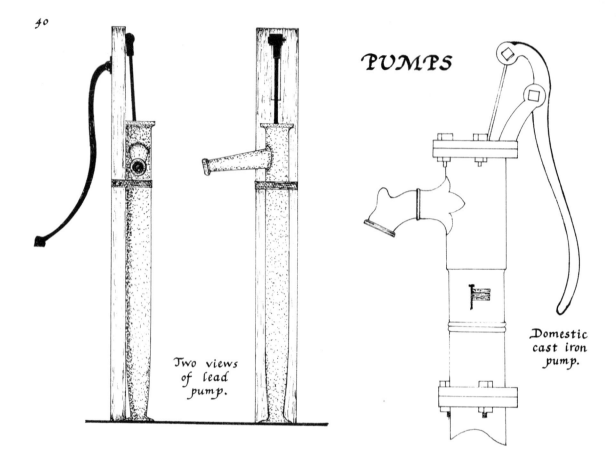

Two views
of lead
pump.

Domestic
cast iron
pump.

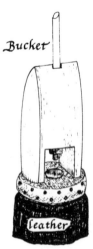

Bucket

leather

    Lead pumps like the one shown above remained in use for many decades. Its barrel is 50 ins. high and has a 3 inch bore. The bucket with its leather skirt rarely needed replacement. Health risks associated with lead pipes were not recognised until recent times. An important development in the C19 was the cast iron pump. It was first used at the top of the well, but it did not take long for plumbers to grasp the idea of extending the well pipe so that the pump could be brought indoors. The domestic offices were transformed by such luxury, and the kitchen maids were at last released from the irksome task of unfreezing an outside pump in the bitter darkness of a winter's morning. The design of the iron pump was simple, but like most plain things it was efficient and robust.

WELLS

The domestic well remained in use until very recent times. Cottage wells often had a simple windlass protected by a tile roof. Galvanised well buckets with their special shape still survive. Some had an eye at the top of the handle to take the hook fixed to the well rope. Shallow wells did not need winding gear. You simply dipped the bucket in on the end of a well hook or stick ~ hence the name stick well.

Communal wells were common in the C19. They often have attractive well houses, which were a focus of social life in the days before piped water. Some readers will recall the times before the 'mains' came ~ for many people this was as recent as the 1940s.

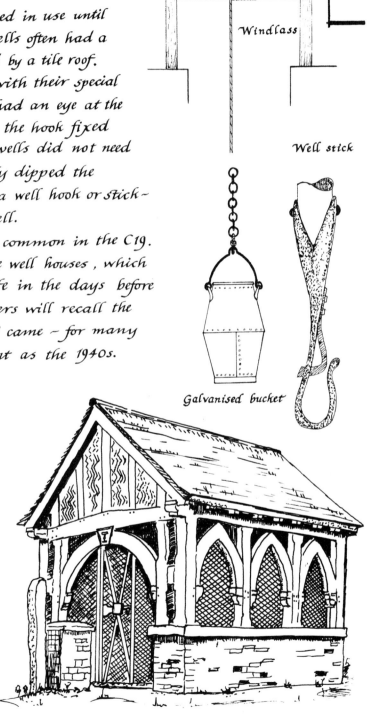

Windlass

Well stick

Galvanised bucket

Grapple hook to recover lost buckets

WELL HOUSE
at
BROUGHTON,
Hants.

bee bole

double bee bole

bee house

skeps

BEEKEEPING

Bees were an important source of sugar in the mediaeval period. Old style hives, skeps, were made of straw ropes bound together in a spiral fashion. There were two shapes ~ flat and domed. Entrance holes were at the top, halfway up the side or at the base. Straw skeps were kept on flat bases of stone or wood which had a landing platform for the inmates.

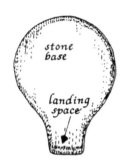

stone base

landing space

To keep hives dry they were given a hat of straw like the one shown below. In many places bee boles ~square recesses in a wall~ can still be seen. They protected the skep from the ravages of the weather. Boles were also made in the gables of houses, and some cottages had bee houses added to their southern walls. One function of the bole was to keep the hive warm.

thatched skep

The custom of 'telling the bees' about family affairs was part of bee lore. Ritual required a teller to take a key on a ribbon whenever a confidence was dispensed.

hangers

nail drawer

knife

ham pot

scraper

bucker

PIG MEAT

pig sticker

chopper

For the cottager and farmer alike the pig was an important element in their diet. The sticking, killing, was usually performed by an itinerant specialist. Thomas Hardy & Flora Thompson (op. cit) have left us vivid accounts of the gory business of sticking. After the bristles had been singed off in a straw fire the carcass was placed on a pig bench to bleed. The blood made the black puddings much prized by those with a taste for such delicacies. After draining the pig was hung on a hanger or bucker. Half sides were treated in a lead salt with a saline solution fortified with saltpetre. After this was completed the muslin cover was stitched on and the bacon placed on the bacon rack near the kitchen ceiling, or hung in the chimney.

scraper made from scythe

hog pot for lard

Nothing from the pig was wasted: the head made brawn ; the chitterlins puddings; scraps finely chopped became sausages; trotters were boiled and the fat was rendered down into lard. The remnants of the larding process made crinklings that were also highly regarded. Hams were cured in the ham pot.

bacon hook

lead salt~ each side was turned & salted daily

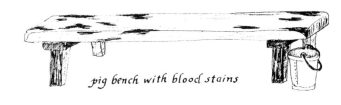

pig bench with blood stains

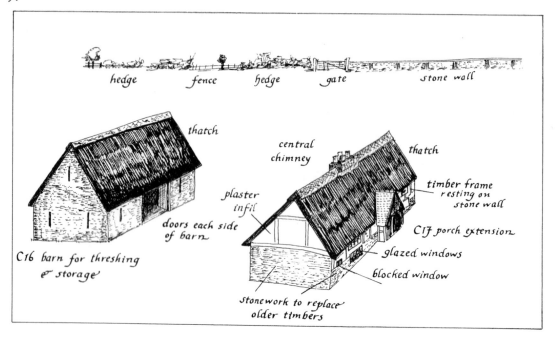

hedge   fence   hedge   gate   stone wall

thatch

central chimney   thatch

plaster infil

timber frame resting on stone wall

doors each side of barn

C17 porch extension

glazed windows

blocked window

C16 barn for threshing & storage

Stonework to replace older timbers

THE FARMYARD

Before the enclosures most husbandry aimed at self-sufficiency. The holders of strips in the open fields were concerned with providing for the family unit. Most of the work was performed out of doors. Individuals did not have the capital to build barns for threshing and other tasks. Apart from the durable tithe barns owned by the Church none of the lesser structures which strip holders must have needed now survive. Such buildings were not important enough to be recorded by writers and we can only guess about their probable forms.

When the open field system changed, fewer owners and larger farms made it possible and sensible for permanent barns to be built. The new arrangement was probably based on a single barn and a farmhouse. The immediate area surrounding both was more easily managed if it was fenced, and thus the post mediaeval farm yard began.

As new methods of husbandry developed the farmer began to need more buildings to house stock, for storage, or for special tasks. These extra buildings could be added more economically if they were made as lean-to structures to existing walls. Almost every farm shows evidence of this kind of development. This typical layout shows a lean-to added to the gable-end and long wall of the original house. In the nineteenth century a new crosswing was added to the northern end to provide space for dairying. There are no milking sheds close to the yard. Cows were often milked afield and the milk carried home – a practice known even in the 1950s. The dairy wing, at the cool end of the house, had an upper floor to provide space for the several living-in workers. Pigs were housed away from the dwelling & leeward of the prevailing wind.

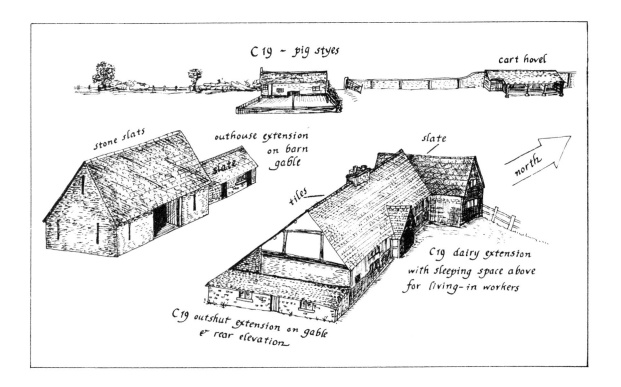

C 19 – pig styes

cart hovel

stone slats

outhouse extension on barn gable

slate

slate

tiles

north

C 19 dairy extension with sleeping space above for living-in workers

C 19 outshut extension on gable & rear elevation

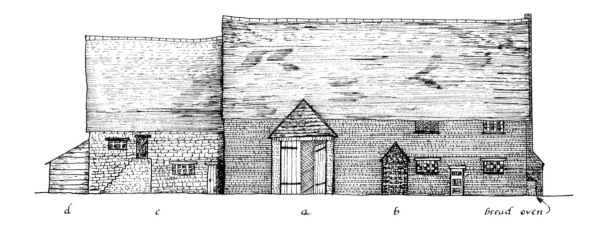

d   c   a   b   bread oven

THE LINEAR FARM

The frequent changes and adaptations which occurred on most
farms often hide the main features of the original farm layout.
There are some farms, however, where the additions can be clearly
identified. This South Midland farmhouse echoes the old longhouse
tradition. It and the barn are contained under the same roof.
Within this C17. brick and tile exterior fragments of a much
older dwelling could be hidden. The window shapes too are C17.
in origin. The high roofline of the main building almost certainly
represents the first phase of this linear farmstead. Two features have obviously been added (a) an extended doorway to
the barn & (b) a brick pantry which, before refrigerators, was
an important part of the domestic offices. At the southern end
the extension (c) provided a stable and storage space above. The
external steps to the upper floor saved space inside. They are a
common feature and may be seen in most areas. A boarded
lean-to (d) has also been added to the principal extension.

This linear arrangement was logistically convenient
for the C17. farmer. When this farmstead was constructed very

few cattle were over-wintered, and there was no need for the kind of buildings shown below. The C19. development of what had been a linear farmstead reflects the changing techniques that agriculture witnessed in the Georgian era. In front of the original buildings a new yard became necessary to provide facilities for increased numbers of cattle and the consequent increase in dairying work. The arrangement of a rectangular yard allowed convenient subdivisions to be made so that straw could be well trodden before it was taken to the fields. Within a yard of this kind space could be provided for all the farm's basic needs; and all the operations were under the eye of the farmer himself.

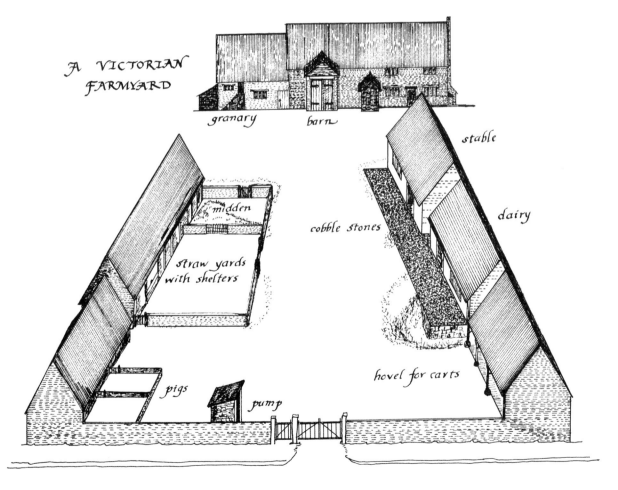

A VICTORIAN FARMYARD

granary

barn

stable

midden

dairy

cobble stones

straw yards with shelters

pigs

hovel for carts

pump

BARNS

One of the farm's most important buildings was its barn. The name is derived from Old English and means barley house. Barley was a principal crop in Saxon England. After each harvest the corn was gathered into the barn for safe keeping and beneath its roof the grain was threshed with flails. Some of our largest and oldest barns date from the C13 and were once owned by the Church. In the mediaeval period the Church could collect its annual tithes, the tenth part of each harvest, which were stored in the vast tithe barns. Tithes in kind were granted to English clergy in A.D. 844 by Ethelwold. They became obsolete in 1836 with the passing of the Tithe Commutation Act (6 & 7 William IV). Then farmers had to erect much more modest structures, & the old barns which will be found on ordinary farms seldom ante date the C17.

Although stone was the natural building material in some places, the erection of a roof always required timber. There are a number of ways of keeping a roof over your head, and the carpenters of old were heirs to an ancient oral tradition that embodied both skill and beauty. A few of the structural designs to be found are drawn below. As economic forces affected the demand for timber so carpenters modified working methods & designs became less robust. Even the lesser woodwork of the C18 looks massive against the pencil-thin timbers used today.

The outside of a barn, if not of stone or brick, is often boarded and tarred - a practical method of preservation. To guess the probable age of a building you need, with permission, to look inside. Timber-framed barns are fascinating subjects for study & each one presents individual features. A timber building is composed of many parts. To allow these to be correctly assembled each was numbered so that the members of a given joint were not mis-matched. Roman numerals, cut with a chisel, were incised on both parts of a joint. These framing marks are usually clearly distinguishable. Some, probably older, buildings bear marks that combine Roman notation with other symbols that are, without doubt, borrowed from the Runic characters of great antiquity. The reason for them surviving

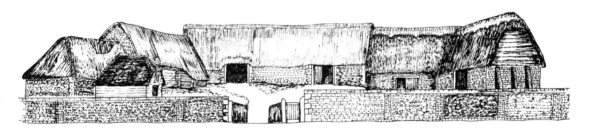

even into the C17 is not easily explained. It is the writer's belief that the signs were borrowed from the mediaeval masons whose 'marks' are to be found in many old churches; & incorporated into the mysteries of the carpenters' craft. The oral tradition could account for them surviving the passage of time. Runes composed of straight lines are easily cut with a chisel.

The size of a barn varied from place to place, but the basic unit of its design - the bay - was a constant. A bay is the distance between each roof truss, & it was determined by the space needed to stall four oxen; which made a plough team. In old measurement a bay was a perch (16 feet) in length. In areas like the southern downlands or East Anglia where corn was a major crop barns with many bays were built. On small farms the threshing barn had at least three bays ~ page 56. Some barns had high hooded portals to allow loaded wagons to enter.

1. Straight cruck. 2. Curved cruck. 3. King post. 4. King post & collar. 5. Braced King post. 6. King post & Queen struts. 7. Queen posts. 8. Braced collar. 9. Braced tie-beam & collar. 10. Braced collar. 11. Tie-beam & vee struts. 12. Braced tie-beam.

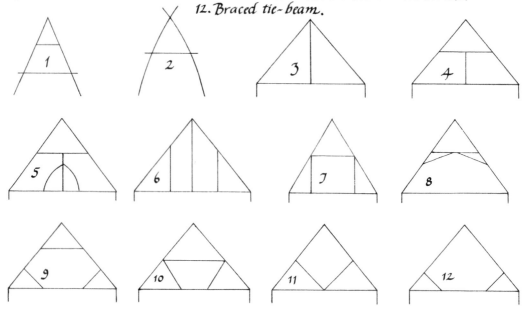

Roof Structure

From the outside most barns look similar. Timber or masonry walls wear a roof of thatch, tile or slate, & you cannot see how it is supported. The diagrams show some types of roof structure. In the Victorian period timbers became slighter & steel brackets & braces were introduced in new buildings. Carpenters of old relied on the volume of timber to give the strength they needed. Joints were pegged together. The tools available to carpenters changed very little from Saxon times until the advent of steam-powered saws. In the days before their invention there was no substitute for the sweat of a man's brow. Our old barns were built with the saw, the chisel, the axe and the adze. Some tie beams were whole trunks barked and roughly squared. Trees were also sawn into sections over a saw pit. Sawyers had a thirsty task. The top sawyer kept the saw in line while the bottom sawyer did the cutting & suffered the sawdust.

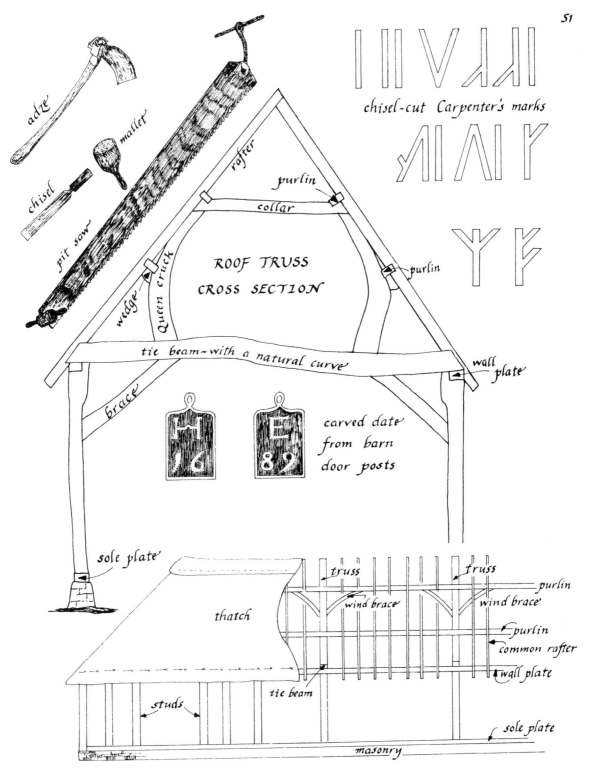

chisel-cut Carpenter's marks

adze

chisel

pit saw

mallet

rafter

purlin

collar

ROOF TRUSS CROSS SECTION

purlin

wedge

Queen cruck

tie beam~with a natural curve

wall plate

brace

carved date from barn door posts

H 16

E 89

sole plate

truss

truss

purlin

wind brace

wind brace

thatch

purlin

common rafter

wall plate

tie beam

studs

sole plate

masonry

SIDE ELEVATION OF BARN

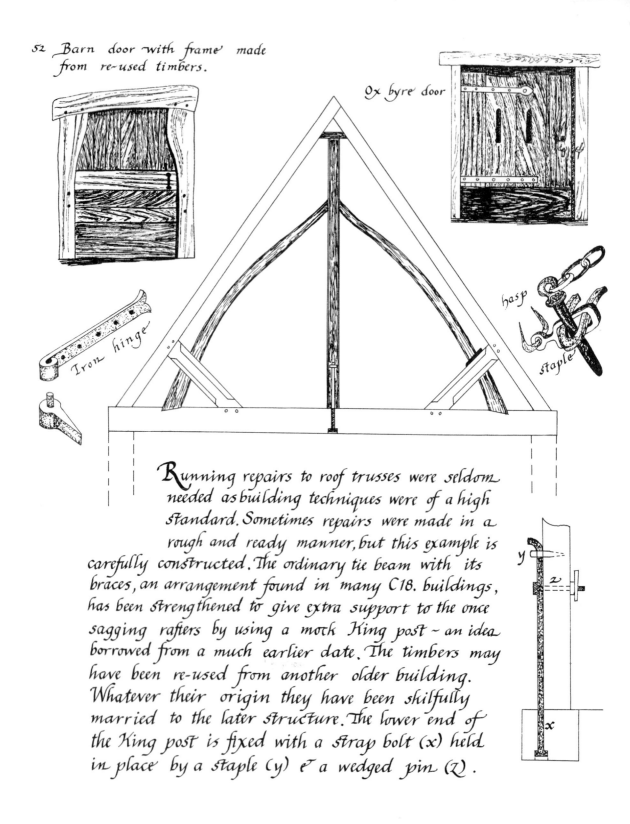

52 Barn door with frame made from re-used timbers.

Ox byre door

hasp

staple

Iron hinge

Running repairs to roof trusses were seldom needed as building techniques were of a high standard. Sometimes repairs were made in a rough and ready manner, but this example is carefully constructed. The ordinary tie beam with its braces, an arrangement found in many C18. buildings, has been strengthened to give extra support to the once sagging rafters by using a mock King post ~ an idea borrowed from a much earlier date. The timbers may have been re-used from another older building. Whatever their origin they have been skilfully married to the later structure. The lower end of the King post is fixed with a strap bolt (x) held in place by a staple (y) & a wedged pin (z).

y

z

x

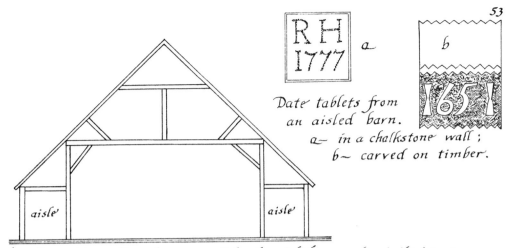

RH 1777 *a* *b*

Date tablets from
an aisled barn.
a— in a chalkstone wall;
b— carved on timber.

England's outstanding barns must be the aisled examples built between C13.& C17. They often have cathedral-like proportions. In the days when all corn was gathered in and not ricked large barns were essential. Many of our remaining aisled barns were formerly tithe barns. At Great Coxwell, Vale of the White Horse, the magnificent barn, once owned by the monks of Beaulieu, Hants., is now in the care of the National Trust. Our ancestors often dated their work. Dates, however, do need careful interpretation. Where more than one date is present the later one may indicate a repair or rebuilding. At the other end of the spectrum there were buildings which had a temporary life of a few years. Few examples of 'bundle thatch' remain, but they were probably important structures to the small farmers of yesteryear. At Manor Farm, Cogges, Oxon. two such buildings have been re-newed. The roofing method was simple and needed a minimum of carpentry. Walls supported a wall plate and a tie beam structure. At the centre heavy branches rested on the tie beams to give support to the rafter-branches, which also rested on the wall. On top of these branches the faggots were piled and trodden into the required profile so that the thatch could be applied.

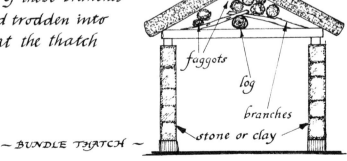

~ BUNDLE THATCH ~

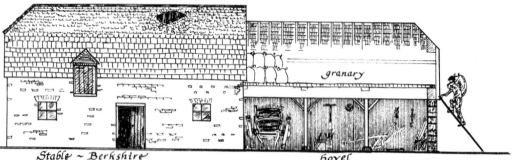

Stable ~ Berkshire   hovel

To the layman most farm buildings are barns. The farmer was more exact.
He distinguished between a real barn and such things as cow houses.
One of the fascinating things about agriculture is the great variety of
regional styles which can be seen in its buildings. The type of farming
practised had a definite influence on the size and layout of the
buildings a particular husbandman would need.
As transport improved in the C19 new building materials began to
appear in areas once the province of tradition alone. On the Berkshire stable cum cartshed, above, the roof is a mixture of stone slats and
imported tiles. The hovel has stone walls but the stable is brick. The
granary was reached with a ladder via a doorway in the gable.
Dual purpose buildings were common. The haybarn below stands on slate
columns and has a slate roof. A lean-to shelter for cattle has been
incorporated on the gable wall. All the materials used came from the
surrounding landscape.

C19 Haybarn ~ Gwynneth
now at
St. Fagan's

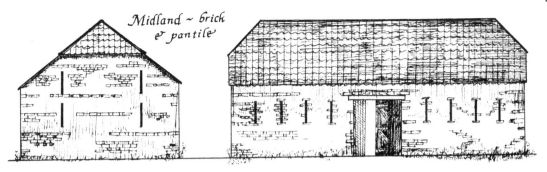

Midland ~ brick & pantile

During the C19, until the depression of the 1870s took effect, large amounts of capital were invested in agriculture. Brick and tile could easily be transported and as a consequence thousands of brick barns were built across the English Midland Plain. Ventilation was often provided by arrow slits placed in the longwall and gable. Pantiles with their cyma curves give a roof a totally different appearance from the stone slats or tiles found in the southern or western shires. These plain barns were sturdy and functional. They mostly survive intact and are a tribute to their builders' skills.

A barn with a slate roof may not entirely derive from the C19. The example below displays two types of roof material. A C16 structure is hidden beneath the weatherboarding. Like most farm buildings it has attracted a number of extensions - a porch, a cart hovel, a hayloft and an outshut shed.

Laithe

Field House

Farm buildings often had particular local names. In Yorkshire the
laithe was a byre, stable and barn. Some laithes had a house attached.
The two elements together were a variation on the long house idea,
and provide us with another example of a linear farmstead.
Another feature of the north is the field house set on its own away
from the farm. It too served a dual purpose with storage for hay
which was fed to cattle in the byre. The dung was then
available to spread on the surrounding land, and so
start the cultivation cycle afresh.

Devon ~ cart
hovel

Granary
& cart hovel ~ Avoncroft Museum

Sussex barn with dovecote

Sheaves	Threshing floor	Straw
mowstead		mowstead
Plan of an Essex barn	door extension	outshut

Essex ~ Three bay barn

Sussex barn with hayloft
and a cart
hovel

The hipped roof and boarded sides of this Sussex barn are typical of the
southern shires. Farmers are adept at tucking things into odd corners
and a small shed has been made in the cart hovel. The simple Devon
hovel opposite was built in the angle of two walls. In the Midlands
a number of cart shelters, like the Avoncroft example opposite, show
us a variation on a common theme.

Drying hops required a kiln. The most picturesque examples, with their
round towers and conical roofs, are those to be found in Kent & Sussex.
In use their white cowls controlled the ventilation. Two or more kilns
were often built together so that they could be linked by a rectangular
block which was used for bagging the dried product. In other places
rectangular kilns were common. Kilns built inside barns
revealed their presence by the cowl on the
roof. At Aldbourne, Wilts. the cowl's
vane is in the shape of a
maltster with his shovel.

Oast House

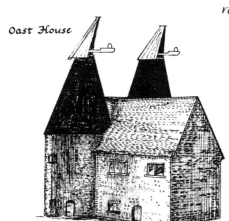

Cowl at Aldbourne Wilts.

swipple

handstaff

For centuries grain was separated from the ear with a flail. Beans too were threshed by hand. There were many local names for the flail and its various parts. A complete flail was called a Stick and a half, but in Brecon it was known as a Joseph and Mary. The part held in the hand was usually called the handstaff. Its shorter part, made from blackthorn, was the fringel or swingel. One of its ingenious parts was the joint at the top of the handstaff which rotated through 360°. On old flails this swipple or swupple (Yorks.) was made from a cleverly carved piece of hazel. The only metal part was the iron pin which linked the joint to the handstaff. The hazel joint was bound together with string. A leather loop on the swingel was bound tightly to it with a thong.

swingle

ferrule

This alternative version has an iron swipple which rests on a ferrule. The loop is fixed with tacks. Flails were still in use in this century.

A flail was a personal tool which was made to suit the stature of its owner.

WINNOWING

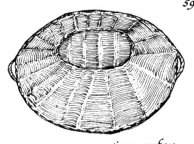

winnow fan

Septvan Arms

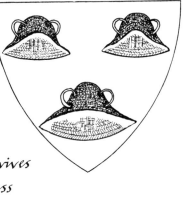

Once the grain had been threshed from the ear it still had to be separated from the husks or chaff - see also p.60 & p.67. The task of removing chaff is called winnowing; a word derived from the Old English 'windwian' - wind. In the mediaeval period winnowing fans were used. These oval wickerwork baskets or fans had two handles. Grain was placed in the basket, and the winnower by shaking the basket up and down caused the grain to be jumped towards the outer rim. The grain remained mixed with the chaff until the draught carried the lighter husks away, leaving the grain in the basket.

A very early illustration of winnowing fans survives at Chartham church, Kent. There the memorial brass of Sir Robert Septvans (1306) portrays him wearing a surcoat over his armour which is decorated with seven (sept) winnowing fans. Mediaeval heralds were fond of puns of this kind. Sir Robert's azure shield of arms bears three gold fans as shown. This use of winnowing fans as a heraldic charge is unique in English heraldry. Two other C15 examples of this shield can be seen on roof bosses in the Great Cloister Vault at Canterbury.

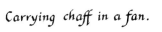

Carrying chaff in a fan.

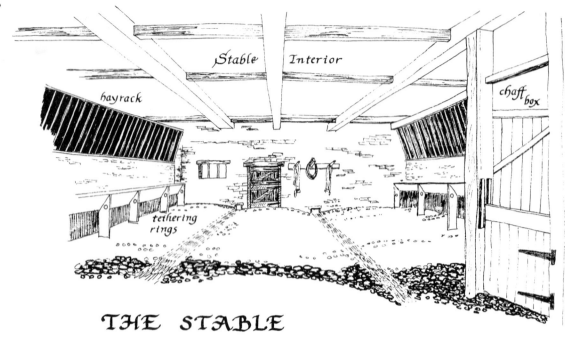

THE STABLE

The horseman's daily round started at 4.0.a.m. so that the horses could be baited (fed) and have time to digest their food before turning out time at 6.30.a.m. As horses worked until 2.30.p.m. their morning feed had to be substantial to carry them through the arduous hours afield. In a stable there was an established order of rank. The head horseman was the first baiter and he always turned out his team before the rest. He appears as the Stabularius in mediaeval records. The bait was often chaff (cut hay & straw) which could be mixed with corn, rolled oats, bran or beans. Hay or straw was cut to make it easier for the horse to eat, & to avoid the wastage that arose from feeding it in its natural length. Stover was the name given to clover hay mixed with straw. Corn was carefully measured out by the head horseman – at the rate of a stone (14 lbs.) a horse a day. It was dispensed with a baiting sieve.

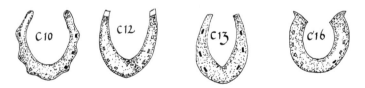

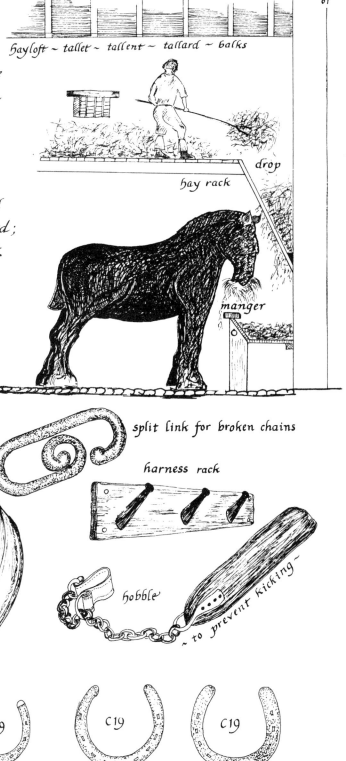

hayloft ~ tallet ~ tallent ~ tallard ~ balks

drop

hay rack

manger

The stable, stelhous in some old documents, usually had a cambered floor so that gullies were formed for drainage. A good stable needed ventilation without undue draughts. Some of the regional names for the hayloft included; balks, tallard, tallet & tallent.

Horseshoes, 'becks' in Kent, changed their shape over the centuries. Heavy horses in Victorian times seem to have had the largest feet.

hoof pick

split link for broken chains

harness rack

baiting sieve

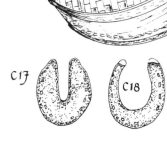

hobble

~ to prevent kicking ~

C17

C18

C19

C19

C19

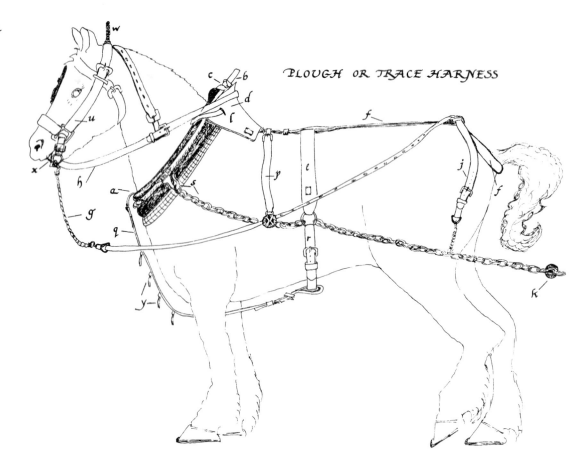

HARNESS

Oxen were the first plough beasts. During the Middle Ages the ox and the horse worked together at the plough, but in modern times the horse replaced the ox in this basic task. Early harness was not made entirely of leather and we know very little about the details of mediaeval horse furnishings. Plough harness had fewer parts than that used for shaftwork. Each piece had a specific job to do. The horse's power was transferred to the plough or cart via the collar and hames. Both fitted round the shoulders where the pulling power resides. Tug hooks on the hames were fixed to the trace chain that was attached to the plough – via the whipple tree. A collar had to be put on upside down and turned after it had been slipped over the horse's head.

The parts of the harness are: a ~ collar, b~ hames, c ~ top strap (it holds the hames together), d~ meeter (which links the crupper or pad to the hames), e~ backstrap (linking the girth and trace chains), f~ crupper (that runs along the back and round the tail), g~ leading rein, h~ hame or check rein, j~ loin straps (fixed to trace or quiler), k~ spreader (to keep traces apart), l~ housen or housing (stops rain running under the collar), m~ saddle resting on the pad, n~ breeching or quiler (fixed to the long staple on the shaft), o~ ridger or ridge-tie (which takes the weight of the shafts to the horse's back), p~ lead guide (with a brass at the end), q~ martingale (joining collar to girth), r~ girth, s~ tug hook and chain (from hames to trace chain or the long staple on shaft), t~ blinker, u~ bridle, v~ saddle cover, w~ swinger or fly terret (worn mostly for special occasions like ploughing matches), x~ bit, y~ brasses (to ward off the evil eye).

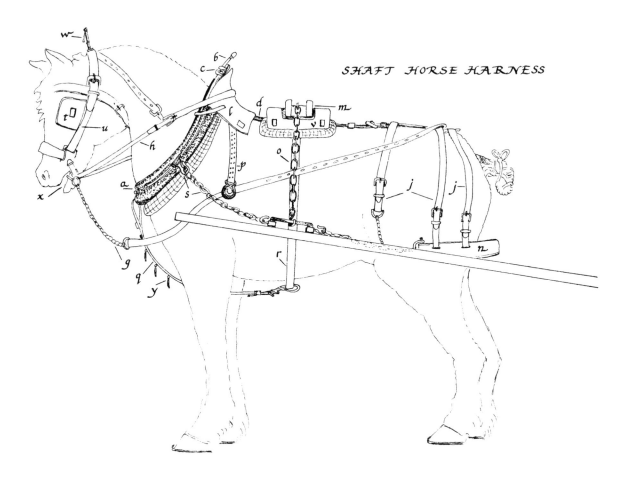

SHAFT HORSE HARNESS

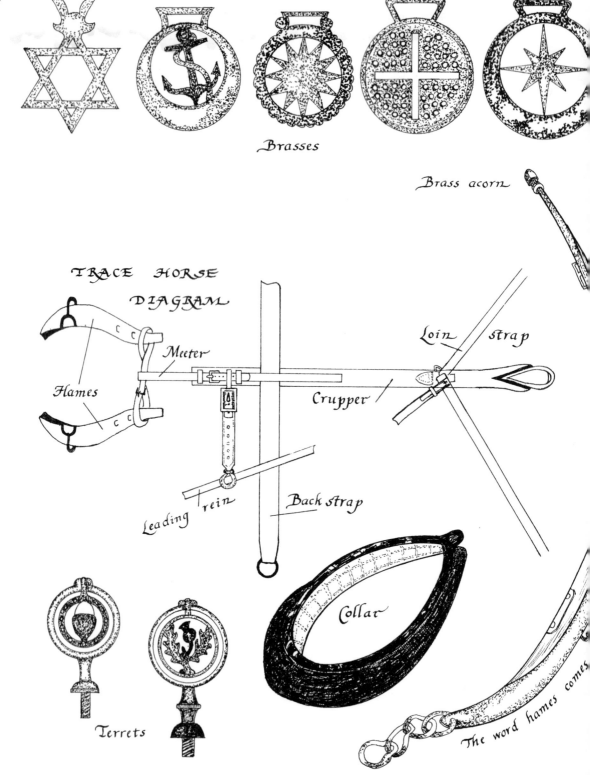

Brasses

Brass acorn

TRACE HORSE DIAGRAM

Meeter

Hames

Loin strap

Crupper

Leading rein

Back strap

Collar

Terrets

The word hames comes

SHAFT HORSE DIAGRAM

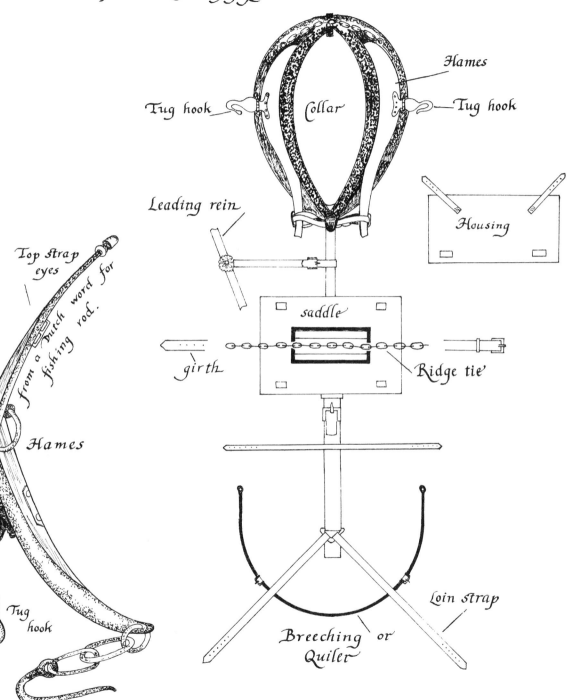

Hames

Tug hook

Collar

Tug hook

Leading rein

Housing

Top strap eyes

from a Dutch word for fishing rod.

Hames

Tug hook

saddle

girth

Ridge tie'

Loin strap

Breeching or Quiler

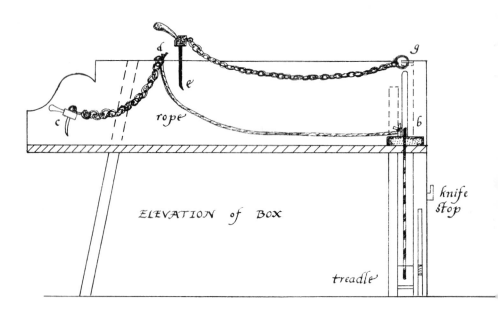

c   *d*   *e*   rope   *g*

ELEVATION of BOX

b

knife
stop

treadle

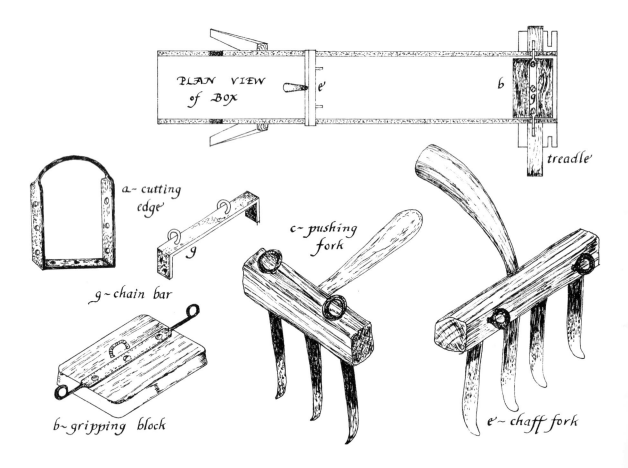

PLAN VIEW
of BOX

e   *b*

treadle

a~ cutting
edge

g~ chain bar

b~ gripping block

c~ pushing
fork

e~ chaff fork

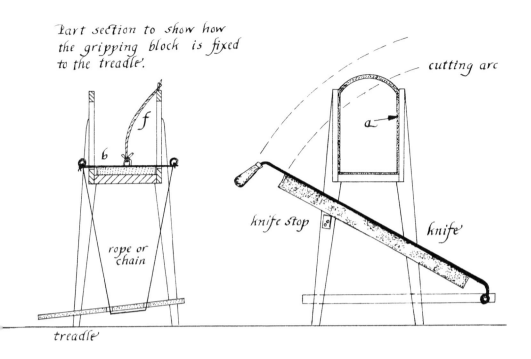

Part section to show how the gripping block is fixed to the treadle.

f

b

rope or chain

treadle

cutting arc

a

knife stop

knife

THE CHAFF HORSE

Chaff cutting was a constant task where there were horses to be fed. The old machines would not pass a safety test by today's standards, and men no doubt lost fingers in the process. This typical machine was probably made by the carpenter. Its ironwork is sparse. The knife could have been made from an old scythe. To give it an edge to cut against the box was fitted with an iron frame (a). A simple treadle pulled the gripping block (b) tight each time a cut was made. This cutter had two operators. One pushed the chaff into the end of the box with a pusher (c) which was fixed to the staple (d). The cutter moved the chaff on with a second fork (e) that slid along the top of the box. After each cut the gripping block was lifted up with the rope (f) which was also fixed to the staple (d). In Kent this type of chaff cutter was called a monkey box or a monkey jumper. The design is old and most likely had its origins in the Georgian period.

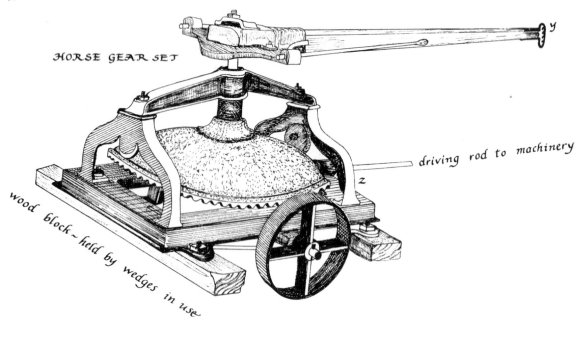

HORSE GEAR SET

driving rod to machinery

z

y

wood block~held by wedges in use

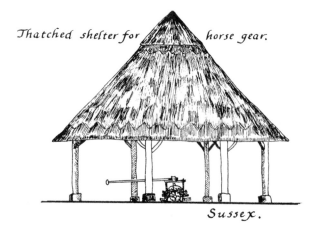

Thatched shelter for horse gear.

Sussex.

HORSEGEAR

Many horse gear buildings were simple structures. Their circular or octagonal plan betrays their original purpose.

19C. extention

18C. barn

GIN GANG

HORSEWORKS or GIN GANG
Northumberland

Winnower, elevator & weighing machine. Hand or horse power could be used.

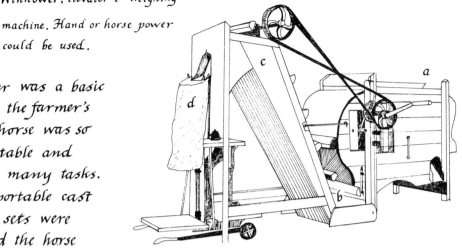

Grain was fed in at (a), discharged at (b) where it was taken by the elevator (c) to the sack (d).

Horsepower was a basic element in the farmer's work. The horse was so very adaptable and performed many tasks. When the portable cast iron gear sets were introduced the horse could work the elevator in the field to speed the making of hay or corn ricks. The same gear could be applied to work in the barn or raising water from a well. Once the gears had been arranged in the required position the horse was equipped with a plough harness ~ collar, hames, backstrap and trace chains with a whipple tree. This latter was attached to the driving arm of the gear (y) and the horse simply walked round in a circle. Power from the gear was taken off by a rod fixed at (z). A series of rods linked with universal joints carried the motion generated into the barn to work the machinery.

Horse gear often became a permanent fixture and then a round house, or gin gang in the north, was built to provide a shelter for the horse and gears. In Sussex and other southern counties this type of shelter was often open-sided and had a thatched roof. In the north stone houses, round or octagonal in plan, with slate roofs were favoured. When other forms of power made these structures obsolete many were allowed to fall into disrepair. Some have survived as store rooms.

CARTING

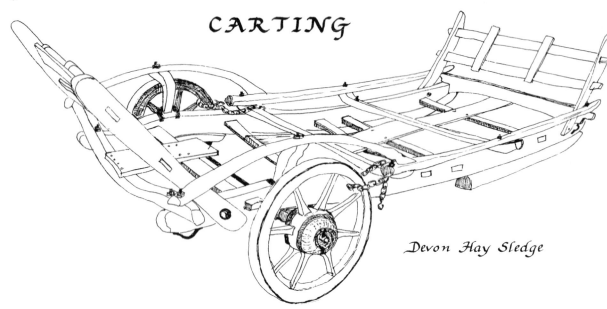

Devon Hay Sledge

Farmers were always moving things around and the manner in which they solved their carting problems depended largely on the nature of the landscape. It seems likely that there were always more two-wheeled vehicles in use than four-wheeled ones. The former are cheaper as well as more useful in confined spaces. There were many variations to the slidecar design on p.131. and in Devon the sledge had its wheels placed at the rear. To add to its capacity ladders were fitted at both ends. The front end of both designs rubbed against the ground. This was an advantage when going downhill with a load.

Butt Cart

On the steep slopes of the small fields in Devon the precious soil was constantly being washed downhill. To maintain the fertility of the field soil was carried uphill again in a small three-wheeled butt cart. The cart was small as soil made a dense and heavy load which could not be moved in large amounts. The cart body tipped up so that the load could be discharged easily.

The maid of all work about the farm was the dung cart. In the C18 carts did not tip-up and their cumbersome build made it necessary for two horses to be used. Lighter carts of the C19 could be drawn by a single horse, although a trace horse could be used if the load or terrain required. In addition to dungwork carts were used at haysel and harvest. Then ladders could be fixed at both ends to support the bulky load. Carts of this type were made by country wheelwrights and in factories. Some bear a maker's plate. Many carts of this type were built in Scotland and sold south of the border. Whatever their provenance the style was often given the label Scotch Cart.

Scotch Cart

THE FARM WAGON

There are few things which demonstrate the uniqueness of our rural traditions more eloquently than the myriad designs we find among our old farm wagons. In such a small country it is a wonder that so many variations on the same theme could be devised; but they were, and our heritage is all the richer for that sturdy independence of mind which made wheelwrights uphold their own regional styles. Two types of traditional design can be distinguished. We call one style a box wagon as it has a more or less flat top to its body. The second has an unmistakable arch above its hind wheels which is properly called a hooped-rave. Within both groups there are many alternative kinds. Almost all farm vehicles had red under-frames and wheels, but the body colours in general use were blue or yellow. Box wagons were widely distributed - from Yorkshire to Devon. Some very old wagons still exist. One of the oldest c. 1795 can be seen in the Rutland Museum.

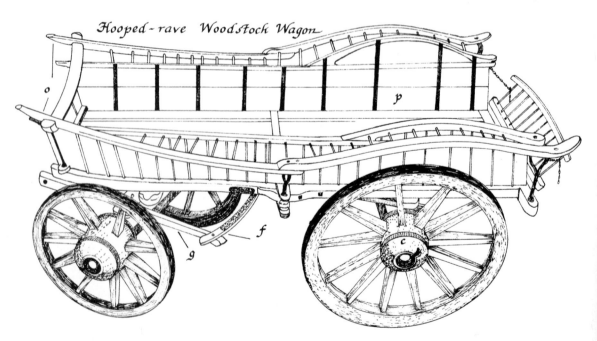

Hooped-rave Woodstock Wagon

South Midlands Box Wagon

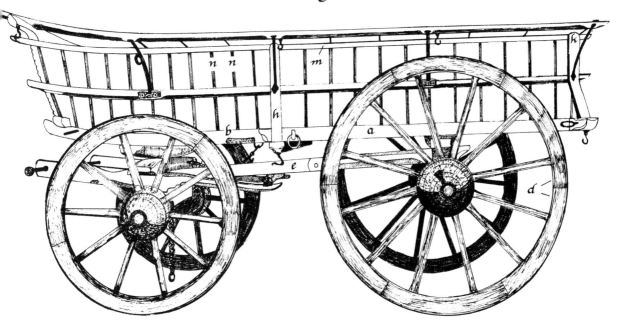

A wagon's front wheels had a turning arc restricted by the wagon
bed (a). To protect it a plate called a locking cleat (b) was fixed at
a point where the wheel came into contact with the side. Wheelwrights had their own language & each part of the wagon had
its own name. A wheel hub was called a nave (c), the rim (d)
was composed of sections called felloes. The front wheels were
linked to the rear axle by a pole (e) that took the rub of the
moving sway bar (f) as the wheels changed direction. The sway
bar was linked to the front axle by hounds (g) & together they
kept the front axle level. To take the pressures of the load the
body sides were buttressed by a wood or iron middle staff (h)
and hindstaff (k). Between the wagon bed (a) and the top rave
or rail (m) a row of spindles (n) gave support to the boards (p).
Another rail, an outrave (o), helped to prevent the load
spilling onto the wheels.

74

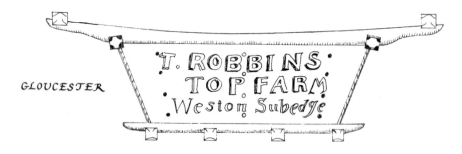

GLOUCESTER

There were many regional varieties of headboard. It was common practice for a farmer's name and address to appear on them. The above example was probably lettered, branded, by a hand unused to setting out such precise details ~ as a study of lines 2 & 3 reveals! T. Robbins's wagon can be seen at the Hereford & Worcester Museum ~ p. 159.

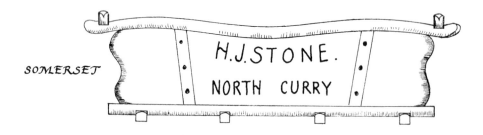

SOMERSET

In Somerset headboards stood in front of the wagon sides. They were also rather shallow and had undulating edges. The lettering on the board above is plain, but symmetrically arranged. A contrasting style was to be found in Kent & Sussex. There a small board fixed above the wagon box, which held the carter's bait and cold tea, gave the essential details.

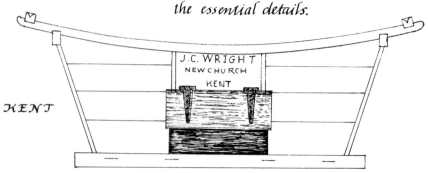

KENT

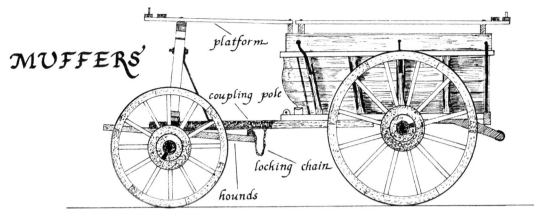

MUFFERS

FORECARRIAGE

Between the cart and the wagon came the 'muffer' or hermaphrodite. This vehicle was a tip cart that became a wagon when its shafts were removed and a coupling pole and forecarriage added. The purpose of this arrangement was to create a flat platform at the front so that greater amounts of hay or corn could be carried. Muffers were unsuited to hilly areas and they worked on the flat lands of the eastern counties. To prevent the forecarriage turning too far below the platform and making the vehicle unstable a locking chain was fixed to the rear of the swaybar.

The forecarriage of the hermaphrodite was composed of the same elements used for a traditional wagon. Thus the hounds were fixed to the axle and the bolster rested upon them. Between the axle and bolster the coupling pole was placed. The pillow supported the sheerlegs and lintel. To give the structure stability it was fixed to the pole by two iron braces. Pillow, bolster and axle were linked together by the king pin that provided the pivot about which the forecarriage turned.

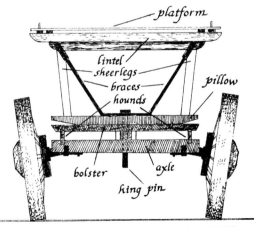

FRONT VIEW OF FORECARRIAGE

tree
pruning
hook

The roller ground apples
to a pomace. Pomace was
crushed in the press to
extract the juice.

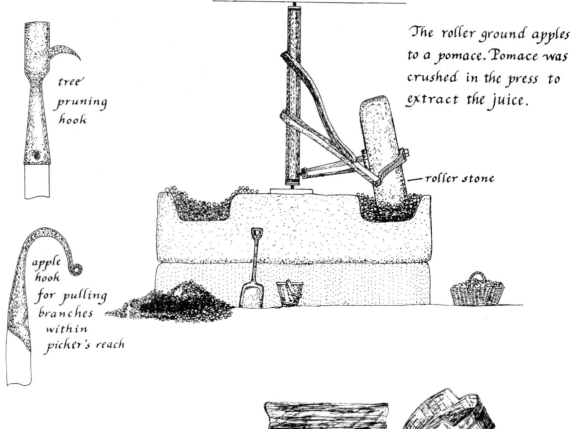

— roller stone

apple
hook
for pulling
branches
within
picker's reach

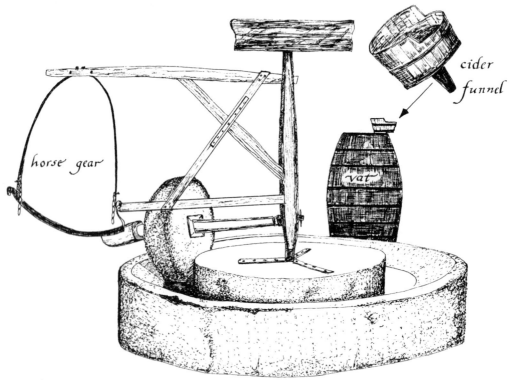

horse gear

cider
funnel

vat

CIDER MAKING

For the countryman, who lived in the western counties, cider was an important product. It went with him into the thirsty fields of haysel and harvest. Making cider was an annual ritual which took place in October & November. Apples heaped in the orchard had to be touched by the frost before they were brought to the mill to be crushed into pomace; by the heavy stone rolling round its deep furrow. Cider mills varied in size but they were usually about six feet across. Pomace was folded into a straw-lined 'cheese' & then squeezed in the press. The rich juice flowed into the stone tray to be collected in the manner shown. To operate the press a bar was placed in the eye of the barrel (e) and slowly turned. After pressing the dried pomace was fed to stock. Following fermentation newly made cider was decanted into another vat. Old brandy or sherry casks were favoured for this purpose as they added to the cider's quality.

On Twelfth Night it was customary to go to the orchards and wassail the trees with rhymes such as ~

Here's to thee, old apple tree,
Whence thou may'st bud, and whence thou may'st blow,
And whence thou may'st bear apples enow!

STADDLE BARNS

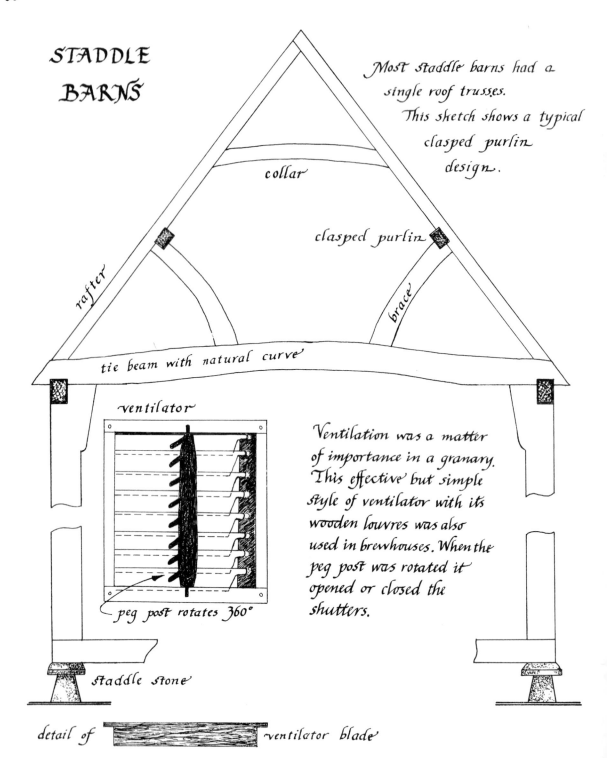

Most staddle barns had a
single roof trusses.
This sketch shows a typical
clasped purlin
design.

collar

clasped purlin

rafter

brace

tie beam with natural curve

ventilator

Ventilation was a matter
of importance in a granary.
This effective but simple
style of ventilator with its
wooden louvres was also
used in brewhouses. When the
peg post was rotated it
opened or closed the
shutters.

peg post rotates 360°

staddle stone

detail of ventilator blade

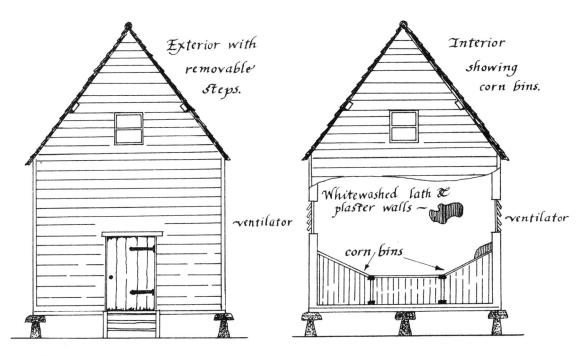

Exterior with removable steps.

Interior showing corn bins.

Whitewashed lath & plaster walls ~

corn bins

ventilator

ventilator

Once the harvest had been gathered the grain had to be stored so that it could be readily available for the miller. We do not know in any detail how the mediaeval strip holder stored his corn, but there is plenty of evidence for the C18. Rodents were the farmer's principal enemy as far as corn was concerned. To keep them out of the corn the granary was mounted on special mushroom-shaped stones which rodents could not climb; they were defeated by the overhang. We call these structures staddle barns. This term comes to us via the Old English word meaning a platform to support a cornrick. Staddle stones are made in two parts. The cap stone can be square or round, but the column is usually a truncated pyramid. Staddles are more often seen these days as garden ornaments.

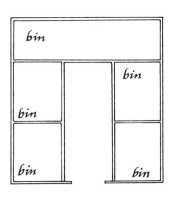

Plan of granary.

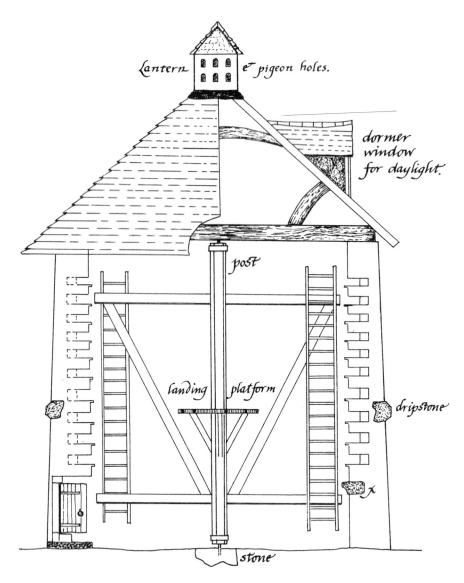

Lantern & pigeon holes.

dormer
window
for daylight.

post

landing platform

dripstone

x

stone

A sectional view of the dovehouse with a potence. The framework
bearing the ladders is supported by the central post. This is fixed
with a pin into the lower side of the tiebeam and also into the
stone placed at the centre of the floor. There is also a landing
platform for the birds. Tiles project from each pigeon hole for the
same purpose. An overhanging course of stonework (x) prevents
rodents climbing up to steal eggs.

DOVECOTES

Noah sent forth a dove, to see if the waters were abated ~ Gen. 8:6.

All over the country you can still see the dovecotes which were very significant buildings for our ancestors. In feudal times only the manorial lord, who was sometimes an abbot, had the privilege of keeping a dovecote. The inescapable need to slaughter most of the live stock before each winter, through lack of fodder, meant that ordinary folk had to make do with salted meat for a large part of the year. Manorial lords had their fishponds, warrens & dovehouses to provide fresh fish & meat during the lean season of the year. Although the birds could eat the peasants' corn the latter did not enjoy the pigeon pies made from the inmates of the dovehouse. Pigeons were kept by the Romans but the noblemen of Plantagenet England are credited with the introduction of the dovehouse to our shires. Most of the early houses were circular in plan. Some of them had a rotating gallows structure, bearing ladders to allow the keeper access to the L-shaped pigeon holes built into the thick walls. Various names are connected with the dovehouse. In Old English the dove was the culfer or culver ~ hence the name culverhouse. From the Middle-English duve we get the dialect form duffus. Many dovehouses were situated close to a pond which gave a supply of drinking water for the birds.

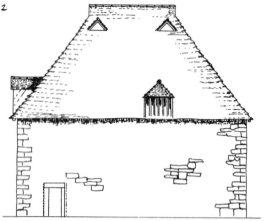

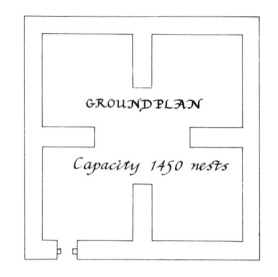

GROUNDPLAN

Capacity 1450 nests

Not all dovecotes were round in plan. This square one with its hipped roof has two tri-angular entrances for the birds just below the ridge. There are two dormer windows to admit light. The door is low by modern standards ~ a reminder that our ancestors were not as tall as we are. From the plan view of this dovecote we can see that there are four short interior walls which add considerably to its total number of nests.

C18. Octagonal

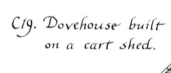

C19. Dovehouse built on a cart shed.

FIELDWORK

STONES troubled the plough as they dulled its cutting edge, and were needed to repair parish roads. From 1555 each parish had to keep its own roads in order. Householders had to give six days' labour a year for this work. Parish highways were controlled by unpaid officers who were known variously as: boonmasters, overseers, stone wardens or waywardens. Their duties remained on the statute book until the Highway Act of 1835. Throughout the C19 children were regularly part of the stone picking labour force. School records refer to pupils being absent for as long as six weeks for this purpose.

In those desperate days of almost perpetual penury for the ordinary farm worker even the pennies earned by the children were significant. Stone picking was often gangwork and no doubt the gangmaster, who contracted with the farmer, kept a strict eye on his workers in the knowledge that— 'One boy is a boy, two boys are half a boy & three boys are no boys at all.'

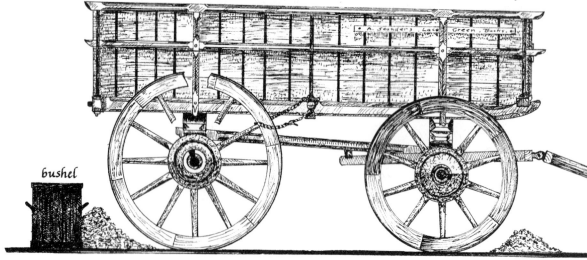

bushel

Picking was usually a job for the winter months between December and February. To spend six or more hours afield in the chill air of a winter's day was not easy work. Payment was made by the bushel. In the 1890s school children could earn about eight pence ~ old style ~ a day. Tools were simple for this elementary task. A stone rake like the one below would save a lot of wear on gloveless hands, but there was no defence against the cold. In the Chilterns a small wagon was used for stone carrying. It had smaller wheels than a normal wagon and its body was therefore closer to the ground ~ an important point for loading. The body was about eight feet long, and its short wheelbase made it easier to turn on hilly ground. A chalk line was often drawn round the inside of the wagon to measure the load.

stone box

stone hammer

stone rake

Drystone
walling

Some of the earliest boundaries probably delineated the extent of
a given parish. Where traditional drystone walls are used (a
technique known by the builders of the Celtic brochs) and they are
contiguous with a parish, the wall itself may be a present sinew
of an ancient past. It is difficult to prove that such a wall has stood
for two millennia, but the possibility that many walls of mediaeval
origin still survive is less far fetched.

When the enclosures came to divide the open landscape into
a patchwork of irregular forms even more boundaries were needed.
In lowland areas earth banks or a post and rail fence served
to divide one field from the next. The C18 saw these boundaries
strengthened by the planting of quickset hedges. In Buckinghamshire a fence was called a mownd, C18. Once the hawthorn hedge had
been developed the craft of hedge laying could begin. In Devon a
hedge was lade & steeped when it was cut and woven together and
then banked up with earth. Mending such a hedge was called
plashing. The running band of hazel which bound the stakes
together was called the ethering.

The C19 saw the advent of wire fences that were cheaper to make
and therefore attractive to the farmer. Barbed wire became common,
& several specific tools, like those shown overleaf, were devised to
strain it tight. Farmers had to allow the passage of travellers
along the footpaths - those most ancient walkways. Stiles were
the easiest way to cross a boundary. They allowed humans to
pass and kept stock confined. On the next page examples of
various stiles are shown.

Zig~zag stile

Glos.

Cambs

Sussex

Wiltshire ~ squeezer stiles

Midlands~ one step stile

step rail

Surrey~ two step stile

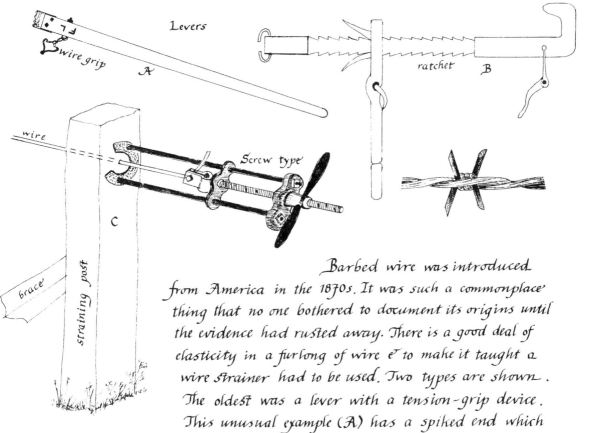

Levers

F.L.

wire grip

A

ratchet

B

wire

Screw type

C

straining post

brace

Barbed wire was introduced
from America in the 1870s. It was such a commonplace
thing that no one bothered to document its origins until
the evidence had rusted away. There is a good deal of
elasticity in a furlong of wire & to make it taught a
wire strainer had to be used. Two types are shown.
The oldest was a lever with a tension-grip device.
This unusual example (A) has a spiked end which
bit into the post when in use. As the lever was pulled downwards
the wire was gripped in the jaws. A more elaborate lever system (B)
had a ratchet to maintain the tension. The screw strainer (C)
was more refined and required less energy from the operator.

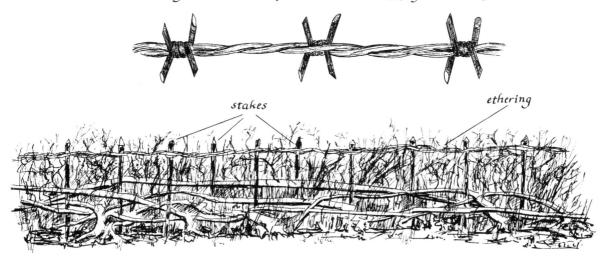

stakes

ethering

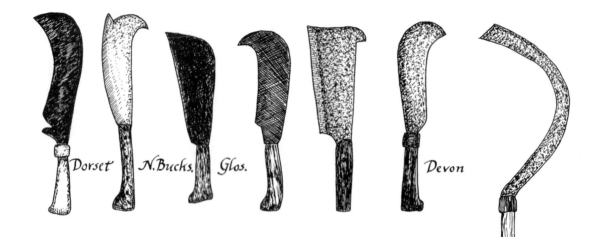

Dorset N. Bucks. Glos. Devon

Slasher

BILLHOOKS

Slasher

The design of the hedger's bill took many forms.
Most of them were made in Sheffield and the
variety of patterns was said to depend upon
local preferences. Thus it was that the catalogu-
es of the various makers attached particular
county names to their special products. From
the practical point of view all the hedger
needed was a well balanced tool which could
be constantly used without unnecessary
effort. Craftsmen are usually inclined to
favour specific tools and it is not unusual
for a man to confine the use of his tools
to himself. A cutting tool is unlike a
spanner and it comes to fit the owner's
hand in a special manner. Names given
to particular shapes by the ironmasters did
not confine their use to narrow geographical
areas. Sussex patterns can be found in

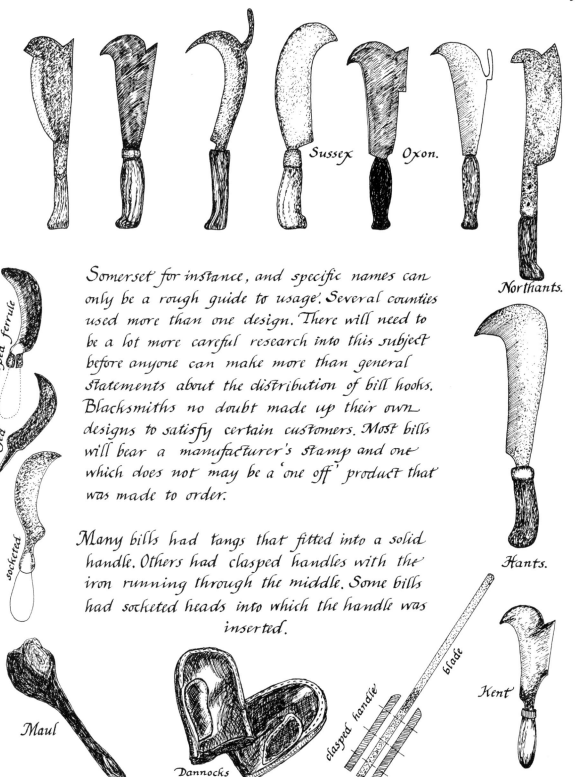

Sussex Oxon.

Northants.

clasped ferrule

tanged

socketed

Somerset for instance, and specific names can only be a rough guide to usage. Several counties used more than one design. There will need to be a lot more careful research into this subject before anyone can make more than general statements about the distribution of bill hooks. Blacksmiths no doubt made up their own designs to satisfy certain customers. Most bills will bear a manufacturer's stamp and one which does not may be a 'one off' product that was made to order.

Many bills had tangs that fitted into a solid handle. Others had clasped handles with the iron running through the middle. Some bills had socketed heads into which the handle was inserted.

Hants.

Maul

Dannocks

clasped handle   blade

Kent

Cross section & line of C18 tapered pipes.

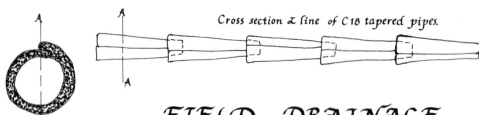

Cretch for compacting bush drains.

Tile pipe layer holding pipe.

FIELD DRAINAGE

Bush drains consisted of hedge trimmings or faggots placed at the bottom of narrow ditches. This style of drain dates from at least the C17. Before the open fields were enclosed it was difficult to arrange comprehensive drainage schemes. In the late C18, tile pipes were introduced. Placed on a sole plate these pipes made quite a good field drain. To see examples of such ordinary but significant parts of our agricultural past you will need to visit your local museum. Another C18. form of pipe was the socketed variety shown above.

Tile pipes were made in country brickfields. The tax which was levied on bricks in 1784 also had to be paid on pipes. In 1826 until 1850, when the tax was abolished, all drain pipes made for agricultural purposes were excused the tax if each pipe was clearly marked with the word DRAIN. Pipes varied in size ~ the usual dimensions ranged from 2 to 6 inches diameter. Drainage is still a costly matter and C19. textbooks show that great importance was placed on removing the least amount of soil. The drain depth varied with the nature and fall of the land. To save

Water filters through soil & into porous pipes.

Cross section of C19 tile pipe.

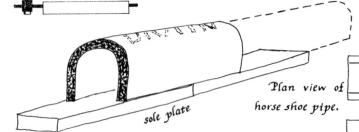

sole plate

Method of setting horse shoe tile pipes.

Plan view of horse shoe pipe.

DRAIN

DRAIN

Stamped sole plate.

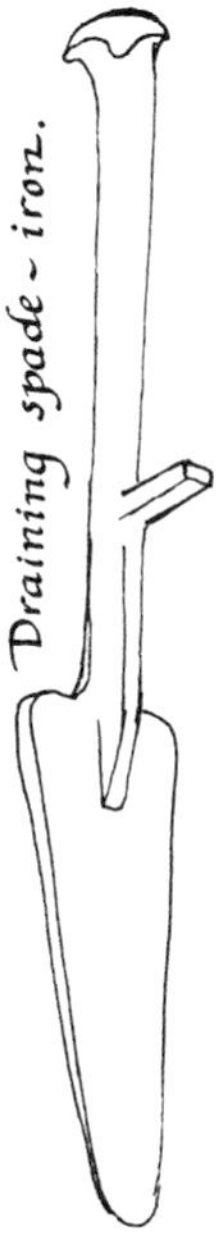

Draining spade ~ iron.

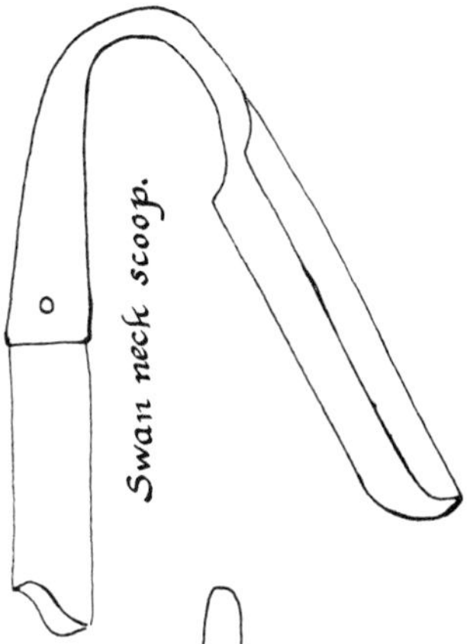

Swan neck scoop.

expense the lower part of a land drain was narrower than the top as the diagram below shows. A wide variety of tools was made for the arduous task of drainlaying. The drain knife was useful for cutting through boggy ground and its blade had a spike which allowed the foot to be used as well. There are several kinds of spade and some are almost entirely of wood. Beech was a favourite choice. Wooden tools were strengthened by metal plates fixed to the front cutting edge or on the back. Some tools like the fly spade are easily identified by their shape as drain tools. The asymmetrical hodder was used for surface work. A leather skirt at the bottom of the slubbing spade's handle helped to contain the sludge scooped from the ditch.

Cross section of drain.

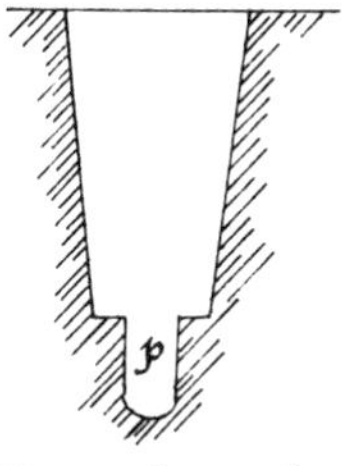

Pipe channel 'p' was made with a swan neck.

Boot with digging iron.

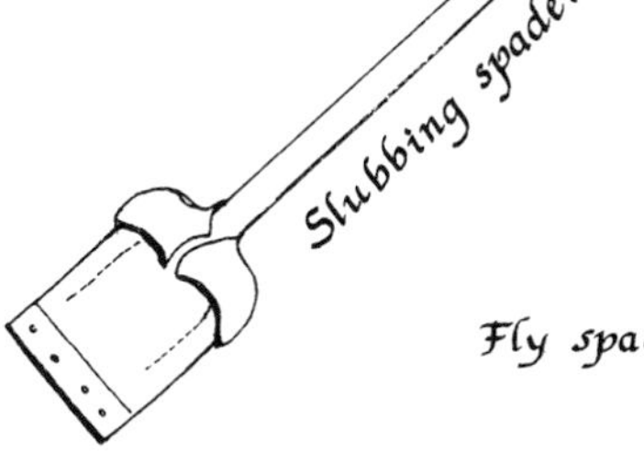

Slubbing spade.

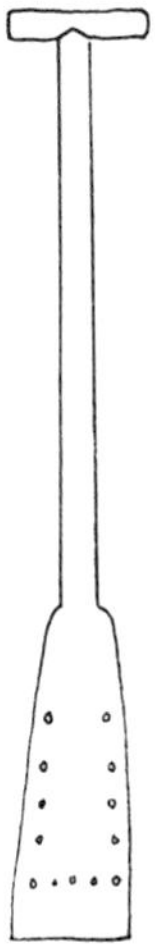

Fly spade.

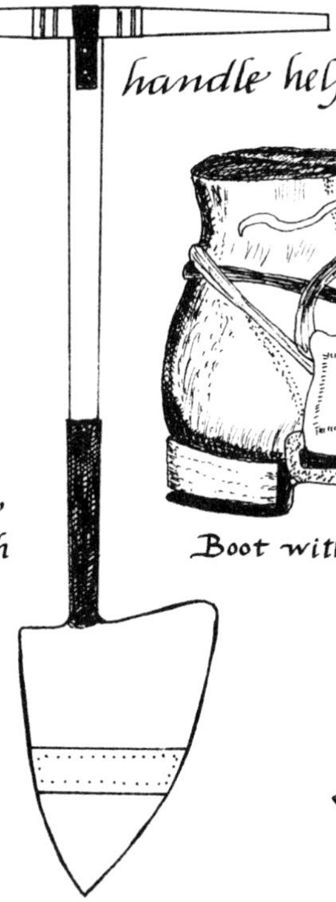

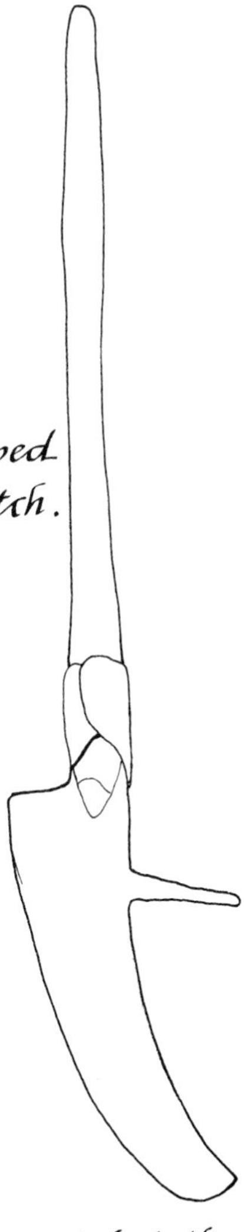

Turf knife.

Hodder or gripping spade.

92

GATES

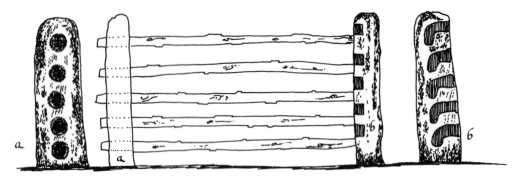

POLEGATE

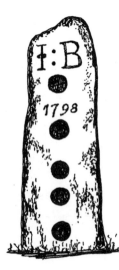

Gates used in prehistoric times were simply hurdles placed across an opening to restrain livestock. Such an arrangement was cumbersome. In the north of England you can still see evidence of the primitive but effective polegate which was composed of five poles & two stone posts. One post (a) had a set of holes with the lower ones closer together ~ a device to deter lambs. The second post (b) had a series of slots. Each pole was pushed into a hole in (a) & then its other end was dropped into a groove of (b). This form of gate was useful as people did not need to remove all the poles in order to pass. A modern version made with horseshoes was still in use in the 1950s & some may still survive. Stone posts may bear dates and initials which make them of special interest to the farm historian.

post & horseshoes

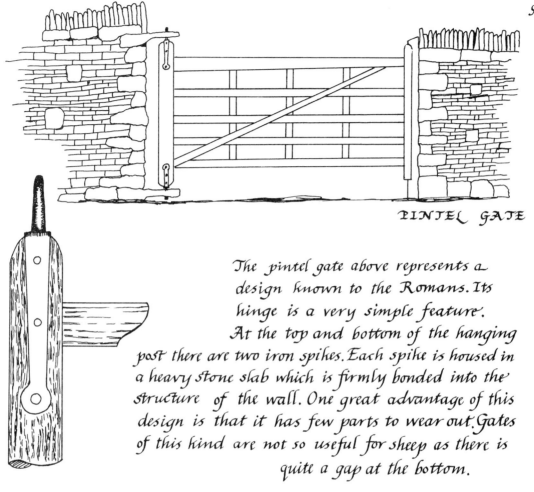

PINTEL GATE

The pintel gate above represents a
design known to the Romans. Its
hinge is a very simple feature.
At the top and bottom of the hanging
post there are two iron spikes. Each spike is housed in
a heavy stone slab which is firmly bonded into the
structure of the wall. One great advantage of this
design is that it has few parts to wear out. Gates
of this kind are not so useful for sheep as there is
quite a gap at the bottom.

Anatomy ~

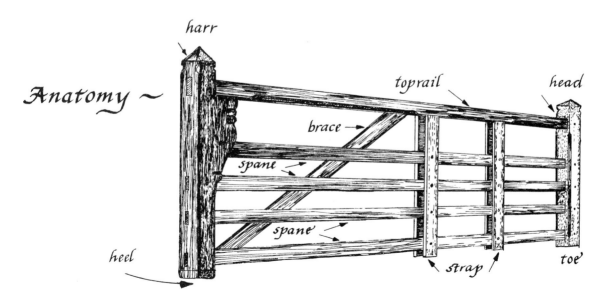

harr, toprail, head, brace, spane, spane, heel, strap, toe

A great deal depends upon the strength of the harr and the top rail. The joint linking these members is the point of greatest stress ~ particularly when the gate is left open. Some designs like the Somerset and the Wiltshire harr are strengthened by an extra thickness. The anatomy diagram shows how the carpenter added suitable decorations to this traditional form. Rails, also called spanes, were placed closer together nearer the ground. Straps were once used on each side of the spane and the nails linking them were clenched ~ bent over. On modern versions of this style of gate single straps are used.

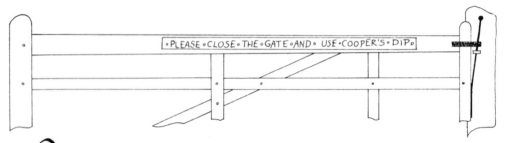

·PLEASE·CLOSE·THE·GATE·AND· USE·COOPER'S·DIP·

The traditional farmgate was wooden with a minimum of iron-work for its hinges and fastenings. Like most things in the count-ryside its precise design once depended upon local tradition. Most gates came from the village carpenter who was the keeper of an oral tradition which had a very long pedigree. Our knowledge of ancient gates is inexact, but the term garstaples, meaning gate fastenings, was in use in the C14. Drawings of gates appeared in mediaeval manuscripts and in early books on farming. How exact the details are is a matter of opinion, but they do seem to be a reliable guide to a gate's main features.

Although many of the old style wooden gates may now have disappeared it is still possible to collect evidence from old photographs and paintings in local museum collections. Sources of this kind can usually be dated. Printed sources such as topographical guides of the 1920s onwards can also supply a good deal of information. The datable photographs they contain often include gateways.

Top rails

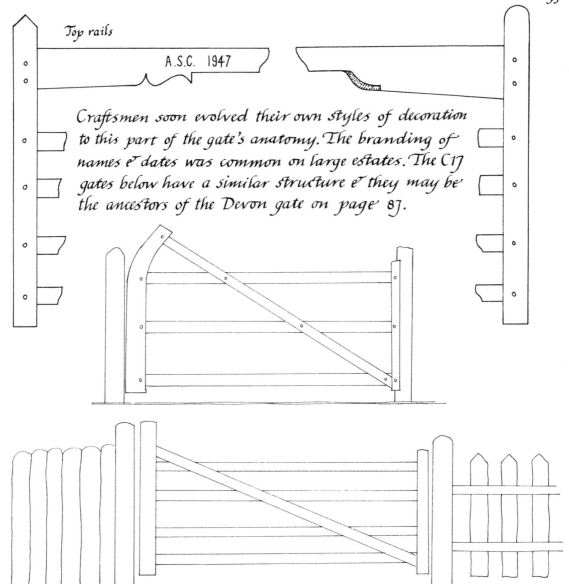

A.S.C. 1947

Craftsmen soon evolved their own styles of decoration to this part of the gate's anatomy. The branding of names & dates was common on large estates. The C17 gates below have a similar structure & they may be the ancestors of the Devon gate on page 87.

Gates are such commonplace things that we take them for granted and seldom give them a second glance. For the farmer a secure gate was and is an essential part of farmstead furniture. The use of wide machines like the combine harvester has caused many gateways to be widened, and the old style wooden gates to be replaced by welded tubular alternatives.

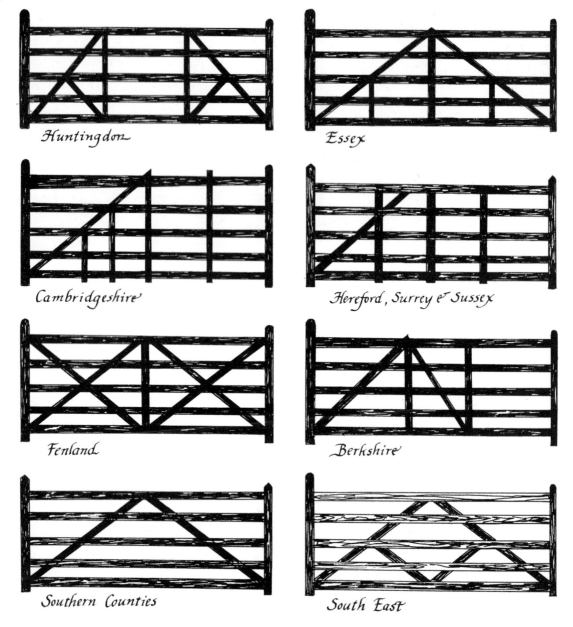

Huntingdon

Essex

Cambridgeshire

Hereford, Surrey & Sussex

Fenland

Berkshire

Southern Counties

South East

There was once a great variety of gate designs to be seen. Local ideas played an important part in deciding the number of braces & straps used. Most fieldgates were 10ft. long and 4ft. high. In 1894 a double braced gate, like the Fenland example, cost about 8 shillings.

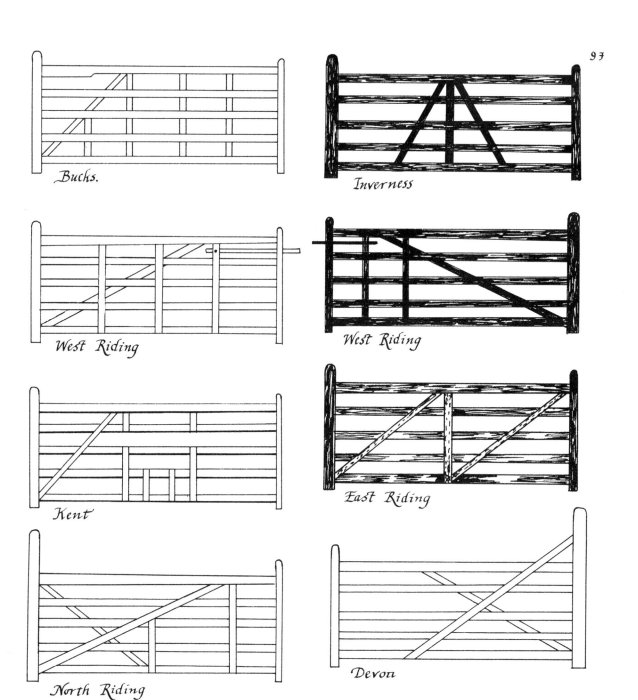

Bucks.

Inverness

West Riding

West Riding

Kent

East Riding

North Riding

Devon

As harrs are always heavier than heads all fieldgates are asymmetrical. Some gates shown here do have a symmetrical arrangement of straps and braces. The West Riding designs use a strap to make a fixing point for a wooden latch.

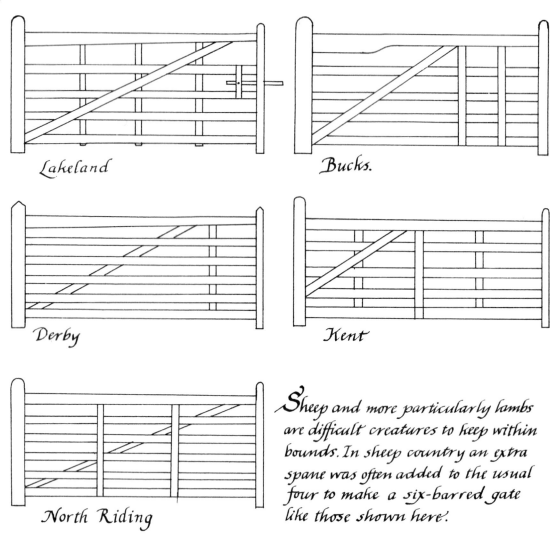

Lakeland

Bucks.

Derby

Kent

North Riding

Sheep and more particularly lambs are difficult creatures to keep within bounds. In sheep country an extra spane was often added to the usual four to make a six-barred gate like those shown here.

The ancient hingeless gate which was bodily moved into place remained in use until recent days. This wide heave gate from Sussex is really a piece of portable fence.

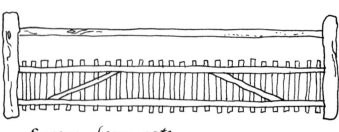

Sussex heave gate

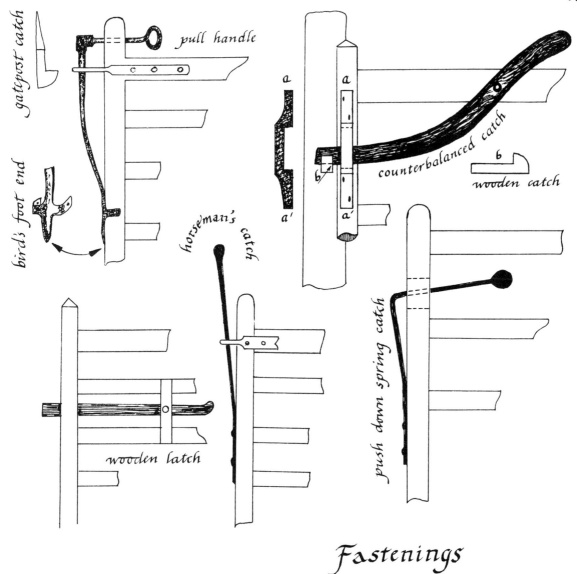

gatepost catch

pull handle

bird's foot end

a

a

counterbalanced catch

b

wooden catch

horseman's catch

a'

a'

b

wooden latch

push down spring catch

Fastenings

There were many different ways of keeping a gate closed. The earliest latches were probably wooden. Village blacksmiths made their own iron fastenings. These simple devices often displayed those touches of skill that distinguish them from the mundane mass produced alternatives. The bird's foot end of the catch drawn above is an example of a craftsman's flourish.

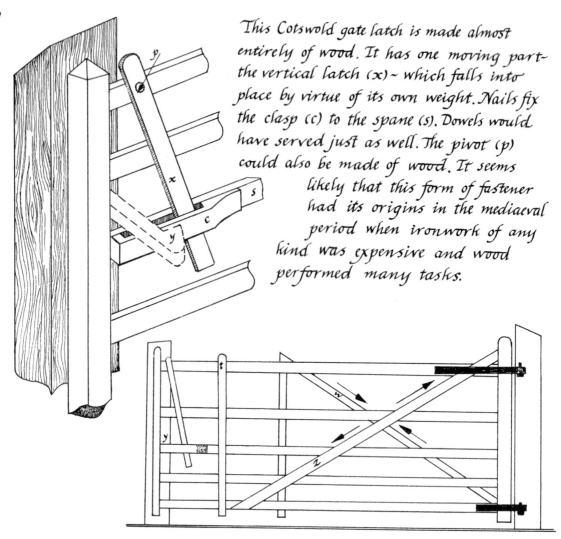

This Cotswold gate latch is made almost entirely of wood. It has one moving part ~ the vertical latch (x) ~ which falls into place by virtue of its own weight. Nails fix the clasp (c) to the spane (s). Dowels would have served just as well. The pivot (p) could also be made of wood. It seems likely that this form of fastener had its origins in the mediaeval period when ironwork of any kind was expensive and wood performed many tasks.

A view to show how the gate is placed so that the catch (y) passes on the inside of the head when closed. There are two braces on the gate ~ one in tension (z) ←——→ and the other (w) in compression —→ ←—. Notice how the latter brace and strap (t) stand proud of the top rail ~ a feature of many designs that has a decorative rather than a functional significance.

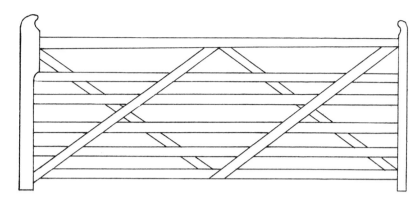

A Monmouth gate which is similar to the Fenland example on page 96, but the braces are arranged in a different manner. There are six bars instead of the usual five. The intersections are pegged which declares a certain antiquity, and the only ironwork is the hinge and a neat but modest spring catch. The harr is interesting and should be compared with the others drawn above. This sketch was one of the last made for this book & the gate, still standing in 1981, could serve for many more years.

Some of our traditional gate designs have been recorded above. The names attached to them do not imply that a particular design is limited to that area alone. As they are such ordinary things they have been neglected. The reader is exhorted to go out and compile a personal collection before all our wooden gates disappear.

A Gloucester gate with an iron brace passing through its horizontal parts to the strap at the toe. Iron through wood is not a good thing as rust decays the timber.

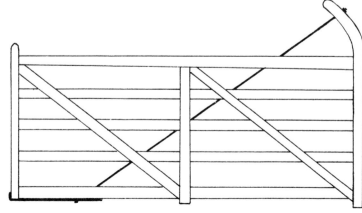

SOIL & SEED

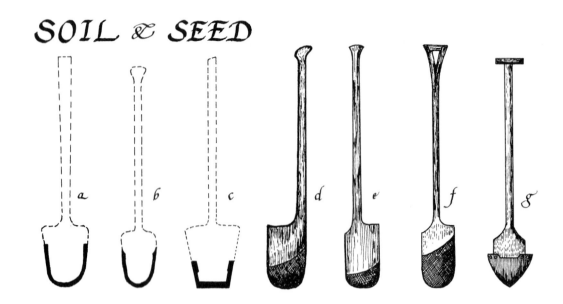

SPADES

Archaeologists tell us that when men were creating the great earthwork at Avebury they used the shoulder blades of sheep for digging. When you begin to examine spade design you have to wonder how the earliest users achieved such perfection so long ago. Roman spades (a–c) are much the same shape as those we use today. These Roman versions were wooden with their working edges protected by iron guards.

Details of Saxon/Norman spades (d–g) come from manuscript sources. The offset blades could be the result of irregular drawing. Side handles could have been an advantage from the digger's point of view as there is more foot room at the top of the blade. There was a plate covering the lower part of the blade which served the same purpose as the Roman guard irons. The last mediaeval tool (g) has a central handle and a shield-shaped metal blade. During the middle ages there was little change in spade design. Tools were made from single pieces of timber re-enforced with iron.

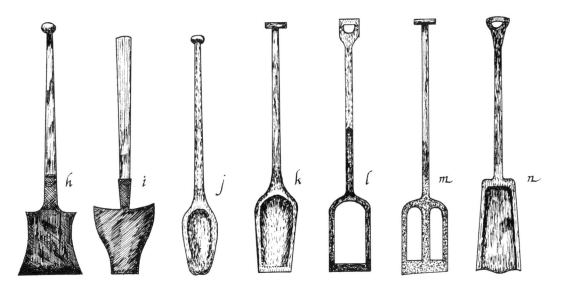

Printed sources allow us to know about the shape of digging tools in the C17. Some handles (h) had a rounded end to make them more comfortable to use. Example (i) made for draining is unusual as it had a socketed head. Wooden spades did not go out of fashion even in the last century. Two examples (j-k) were made for drainage work. Notice the different types of handle. In common with many wooden digging tools the handle becomes wider near the top of the blade where it takes the greatest levering stresses. Heavy ground was always a problem. The spades (l & m) are versions of the open spade used to cut through heavy clay. Spades of this kind offered less resistance on each downward stroke. Specimen (n) is a wooden drainage spade circa 1920.

Prehistoric digging tools: x ~ a shoulder blade shovel with an antler handle; y ~ an antler pick; z ~ a stone hammer used to strike the pick at 's'.

MUCK SPREADING

manure knife

One of the important tasks in the farming year was muckspreading. You had to feed the land. Beasts kept in yards - called tothe folds in Banff - during the winter trod the straw and provided the shairn - dung - that ended up in the midden or mixen. The Warwickshire word was bury, in Suffolk it was a pie clamp and in Hereford a ruck. As the heap grew it became solid enough to need a manure knife before it could be carried away. Manure was known by various names e.g. greeds (Kent), soil or swile (Glos.), till or tillage (Yorks.). Spreading was also called scaling (Yorks.) and mending (Lancs.). Tools too had particular names. A crome or cromb with three downturned tines was widely used for raking out the muck cart. The word comes from Old Frisian via Middle English and indicated a claw or talon. In Scotland a three tined fork was a graip which became an evil (Cornwall), yell (Staffs.) and a yelve or sharvel (Salop). Homemade tools like the wooden fork probably had very ancient origins.

wooden dung fork

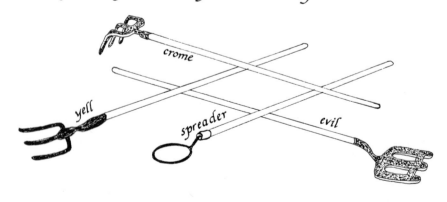

crome

yell

spreader

evil

THE PLOUGH

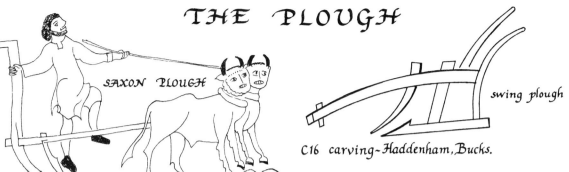

SAXON PLOUGH

swing plough

C16 carving-Haddenham, Bucks.

There are few objects more symbolic of agriculture than the plough. Its origin lies deep in the mists of time. If you have ever heard the rip of the coulter and felt the surge of the share as it drives through the soil you have been in touch with an elemental mystery, that was part of the ploughman's daily experience. On this side of heaven the pulsing plough was among man's earliest implements which shaped the landscape. Old English 'sulh' means plough, & place names like Sulgrave, Sulham and Sulhampstead may derive from it. The word probably meant a gully or narrow valley too. The surnames plowright & plowman as well as the vast number of inns also bear witness to the plough's importance. In prehistory it was a rudimentary device which scratched the ground but did not make a furrow. Several cross ploughings were necessary to form a satisfactory seed bed. The Gaelic cas chrom is related to the later plough form. In use it was pushed into the soil by the user's foot. The earth was then loosened by swinging the handle to and fro.

yoke

beam

ox-bow

IRON AGE PLOUGH

wedge

share

CAS CHROM ~ a foot plough

19C KENT PLOUGH

SECTIONAL VIEW

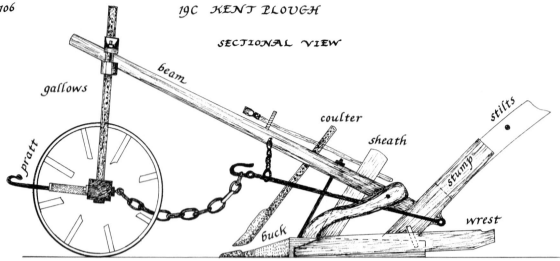

Manuscripts show us how the man operated plough was adapted
to animal power, but the development took place in the prehistoric
period. At Butser Hill ~ see p. 156 ~ the Iron Age Farm Experiment
has advanced our knowledge of ancient ploughing. Ploughs with
wooden shares did not last long and iron must have been used
at an early stage. Archaeological evidence shows us that iron
shares were used in the Romano-British period. Ploughs with
wheels were part of the Saxon scene. The advent of the coulter &
mouldboard marked a significant advance. With a mouldboard the furrow slice could be turned over and also form a
path for the oxen to follow. Long furrows could be laid out and
this saved time in turning about.

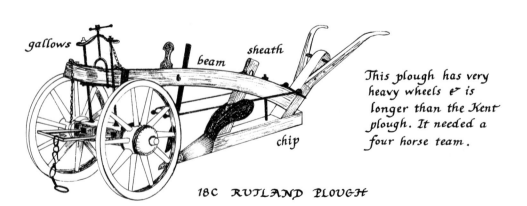

This plough has very
heavy wheels & is
longer than the Kent
plough. It needed a
four horse team.

18C RUTLAND PLOUGH

There was little change in plough design during the Middle Ages, and wooden ploughs continued to dominate the scene. Local usage determined the shape of the members. A wheel-less plough is called a swing plough. In some places wheel-less ploughs had a support at the front of the beam, & they are known as foot ploughs. Ploughs with wheels seem to have co-existed with swing ploughs for a considerable time. The choice partly depended upon the nature of the soil. A Kent plough was a heavy implement but it was suited to a specific terrain. Variations of it were used from Hampshire to the west and as far afield as Rutland. The Kent plough had an important advantage in its moveable mouldboard - also called a wrest, wreest, riste or rice. A fixed mouldboard will turn the furrow in one direction and such a plough had to be worked round the field and not across it. The Kent plough's wrest allowed it to work across and the furrows were turned on top of one another like this
On hilly ground this was an advantage.

To plough quickly and effectively both coulter and share had to remain sharp. Ploughmen always had difficulty in keeping a keen edge on their ploughs and if a plough was blunt it produced ragged work for all the world to see.

WILTSHIRE FOOT PLOUGH
~ after Tom Hennell ~ op. cit.

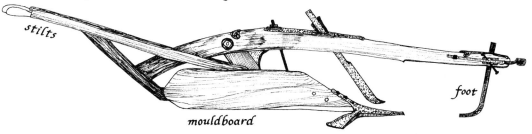

stilts

mouldboard

share ~ wrought iron

foot

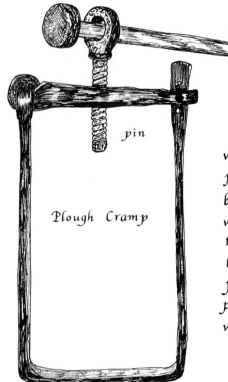

pin

Plough Cramp

Wooden ploughs were armed with various plates and straps. In order to fit them they had to be held firmly in place. The village carpenter, who was often the plowright too, had the ironwork made by the blacksmith. Once the ironwork was secured by the plough cramp the heavy nails could be driven home. A plough's wearing parts are few. The durability of the old style ploughs is demonstrated by those which survive in country collections.

Buckinghamshire Swing Plough ~ C used at North Marston.

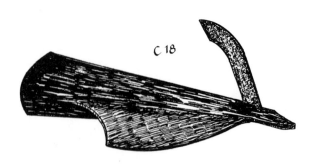

C 18

C 17

Ploughshare design changed as new ploughs were developed. Very early shares sometimes had fin coulters like the examples above. The C18 specimen is 18 inches in length. Its fin is 8 inches high.

Ploughmen took great pride in their work knowing that each furrow would attract the critical eye of their peers. The self-sharpening share was patented in 1785. Robert Ransome perfected a way of tempering cast iron shares so that the base of the share was harder than its upper surface. The softer upper part wore away and thus kept the edge constantly keen. Ploughmen soon appreciated the superiority of cast shares over the wrought ones. Ransomes also developed the idea, in 1808, of interchangable parts which made it possible for a standard plough frame to be adapted for use on any type of ground. The age of mass produced ploughs had dawned. Early in the C19 the plough made entirely of iron appeared, but wooden ploughs endured on some farms into the present century. During the Victorian era a hybrid plough with a wooden beam and stilts, but an iron mouldboard and other fittings was

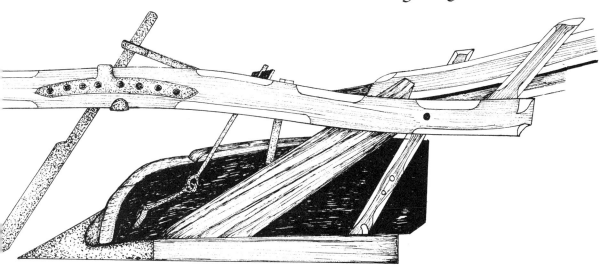

produced by several manufacturers. Steam powered engines could haul ploughs which turned six furrows at a time. These giant machines had an impact on farming, but the small farmer carried on in his own way and untouched by them left progress to fend for itself.

SWING PLOUGHS

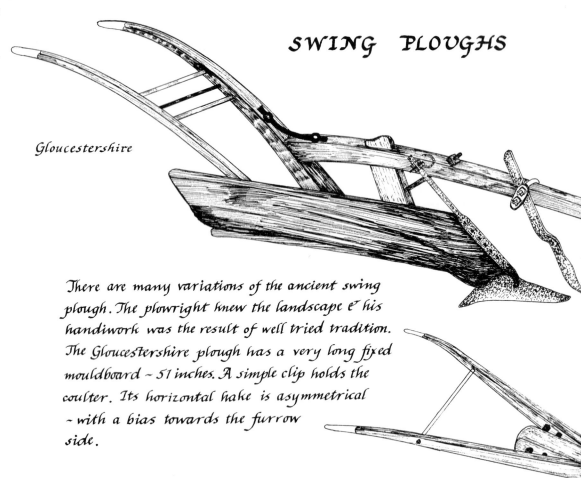

Gloucestershire

There are many variations of the ancient swing
plough. The plowright knew the landscape & his
handiwork was the result of well tried tradition.
The Gloucestershire plough has a very long fixed
mouldboard ~ 51 inches. A simple clip holds the
coulter. Its horizontal hake is asymmetrical
~ with a bias towards the furrow
side.

The Sussex Foot Plough provides a contrast to the Gloucester
example. Its chisel edged share should be compared with
the others. In the lower sketch the position of the shallow
mouldboard with its sideways curve can be seen.
This mouldboard could be moved to the other side
of the plough so that all the furrow slices fell in the
same direction. Then the plough staff was moved to the
other side of the coulter which moved its point to
the opposite side of the share. This plough has a symmetrical
hake. The foot is adjustable.

There were many different names of local importance for the
plough's various parts.

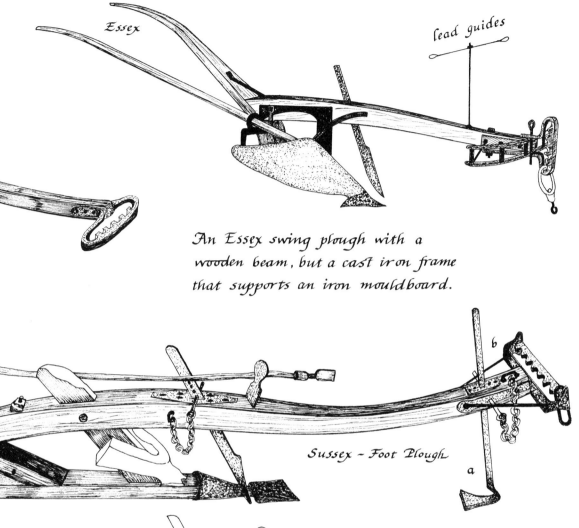

Essex

lead guides

An Essex swing plough with a
wooden beam, but a cast iron frame
that supports an iron mouldboard.

Sussex ~ Foot Plough

a ~ foot. b ~ hake or pratt.
c ~ coulter. d ~ plough
staff or roadbat. e ~ chip.
f ~ point or share.
g ~ buck. h ~ hog.
i ~ wrist, wrest.
j ~ stump. k ~ sheath.
l ~ wrist pin. m ~ stilts.

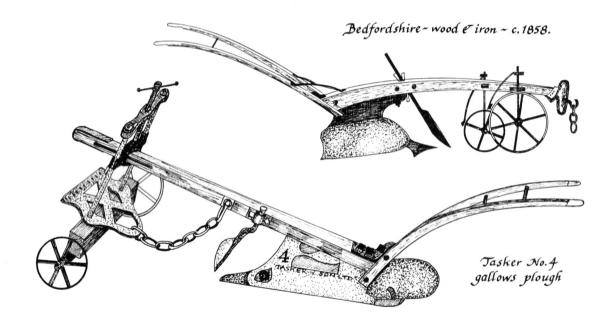

Turning a furrow.

Bedfordshire ~ wood & iron ~ c. 1858.

*Tasker No. 4
gallows plough*

*A skim plough ~ C19 ~
which pared the surface
like a breast plough:
see p. 144.*

SPANNERS

Each manufacturer provided spanners so that the ploughman or reaper working afield could carry out necessary adjustments. Very little has been written about these tools which became redundant when the machines they served were made obsolete by progress. Old spanners were made for square nuts and the shape of a particular tool usually tells you if it was intended for them. Makers' names do not always appear on the tools they supplied. Where they are shown they do give us clues to a tool's possible use. Most spanners had a hard life and they often show considerable signs of wear. Look out for the burrs which come from hammering. Such were the forces applied to spanners, many became twisted like the specimen of the Deering hayrake spanner drawn above. The ferocity of the hammering caused one end to break.

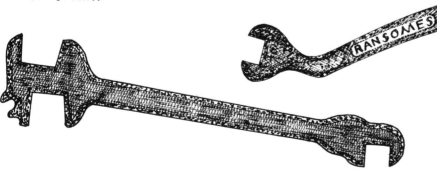

114

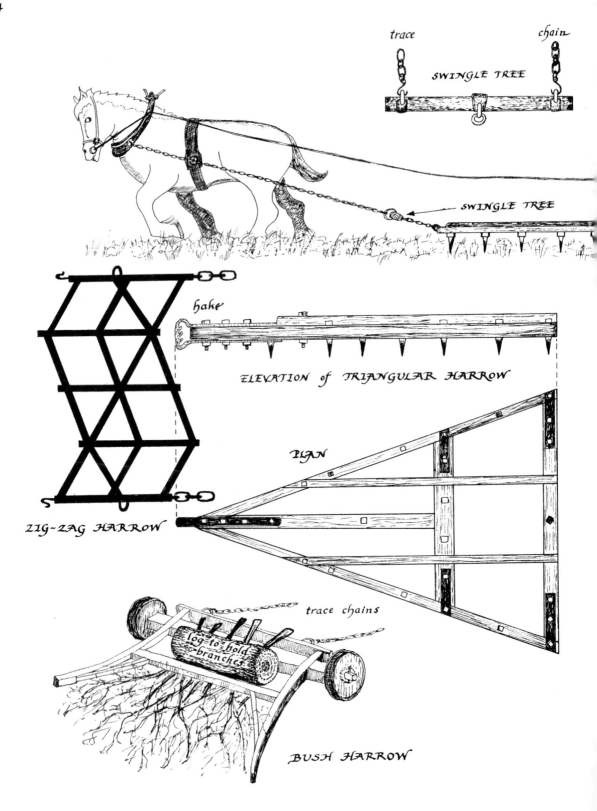

trace

chain

SWINGLE TREE

SWINGLE TREE

hake

ELEVATION of TRIANGULAR HARROW

PLAN

ZIG-ZAG HARROW

trace chains

log to hold branches

BUSH HARROW

HARROWING

beetle

After the ploughing was done the task of preparing the seed bed continued with the harrow which reduced the furrows to a flat surface. In mediaeval days & until the C17 larger clods were broken with a beetle. Harrows were used long ago as manuscripts like the Lutterell Psalter show. They were made by the carpenter and blacksmith. The latter made the hake and tines. A cheaper form of harrow remained in use until the C19. It was made from branches held together in a rude wooden frame, and known as a bush harrow. Harrows varied in size. Pairs were linked together by chains & hooks. Wooden rectangles and triangles were traditional shapes but zig-zag iron harrows were probably more effective. Gangs of iron harrows could cover a wide area at each traverse. The wooden example below has matching curved sides, about 6ft. long, which were formed from the same tree. Its stock was divided into four members which, placed together in opposite pairs, gave the harrow its symmetry. It was pulled in the direction shown. The tines thus made more scratch marks and did not follow each other in five precise rows. Horsepower was linked to the harrow by a trace chain fixed to the horse's collar and the swingle tree.

direction of travel

scratch lines

HANDSOWING

Sowing by hand is older than the Old Testament and it was still practised in some parts even in the C20. Labourers used this method, along with dibbling ~ see page 124 ~ on their precious allotments. Seed was sometimes carried in an apron but seedlips, like the ones drawn here, were in general use. Some broadcasters used two hands for sowing. Styles of seed lips or boxes varied from place, and they included rectangular, and rather heavy, boxes as well as the curved specimens shown. In Wales seed lips were also made with straw ropes in the same way as the beehives on page 42. Straw lips were pliable in use and could assume a profile which matched the sower's body. A tray type was used in East Anglia. It was made with a bent wooden rim which has a canvas or cloth lining.

broadcasting

seedbox

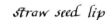

seed tray

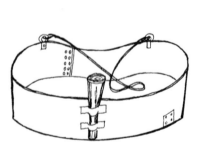

seedbox or lip

seedbox or lip

straw seed lip

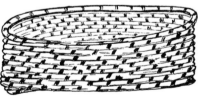

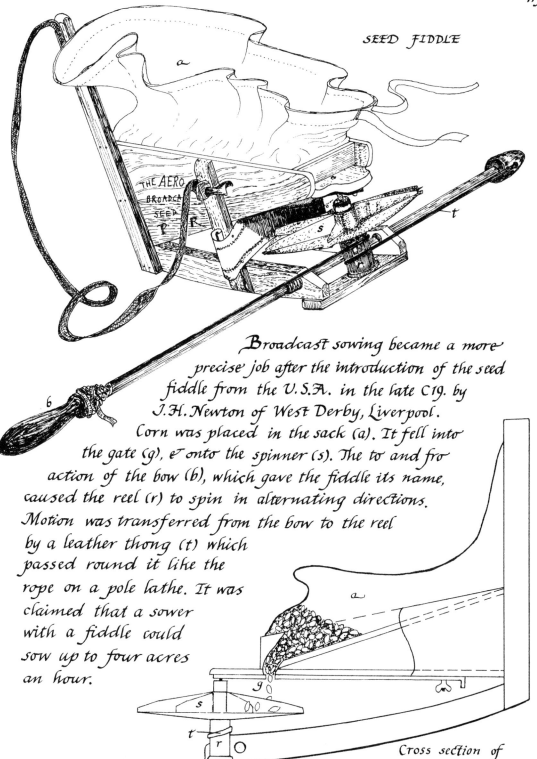

SEED FIDDLE

Cross section of
FIDDLE

Broadcast sowing became a more precise job after the introduction of the seed fiddle from the U.S.A. in the late C19. by J. H. Newton of West Derby, Liverpool.

Corn was placed in the sack (a). It fell into the gate (g), & onto the spinner (s). The to and fro action of the bow (b), which gave the fiddle its name, caused the reel (r) to spin in alternating directions. Motion was transferred from the bow to the reel by a leather thong (t) which passed round it like the rope on a pole lathe. It was claimed that a sower with a fiddle could sow up to four acres an hour.

SEED BARROWS

Seed barrows of this kind were made in large numbers in Victorian days. Although this example has a single handle, two handled versions were equally common. For collectors small machines are attractive items as they can be restored with simple tools. Even if the seedbox is wormeaten it can be remade by using the original as a pattern. A maker's plate was usually fixed to the lid. This normally included a place name which can be useful evidence as it allows one to refer to old directories for further information. The editions of Kelly's county directories published post 1860 contain useful details about agricultural implement makers. A machine's place of origin is called its PROVENANCE. Collectors should try to record as much detail as possible about their acquisitions.

SIDE VIEW of BARROW

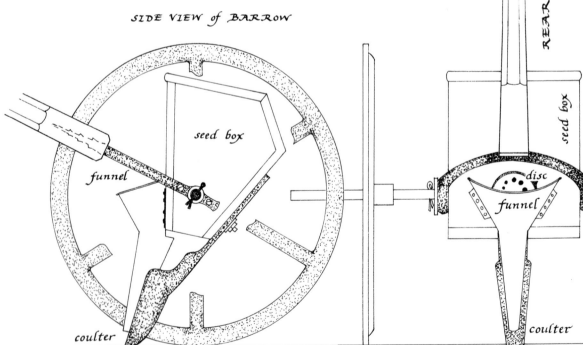

REAR VIEW of SEED BARROW with adjustable wheel spacing.

seed box

funnel

coulter

disc

funnel

coulter

Broadcasting by hand was an inefficient manner of sowing seed. It had served man for several millenia before Jethro Tull found a more exact way with his seed drill ~ 1701. Grain was moved from the seed box in a very simple way. A brush moves at the same speed as the wheels. The motion of the brush pushes the seed towards the back of the box, and each seed is expelled through a hole. It then falls into the funnel and down to the ground where it rests in the drill - a small furrow ~ made by the coulter. The feet of the operator could then tread the soil into the drill to cover the grain. At the rear of the seed box a disc bearing several different holes can be adjusted to suit the crop being sown.

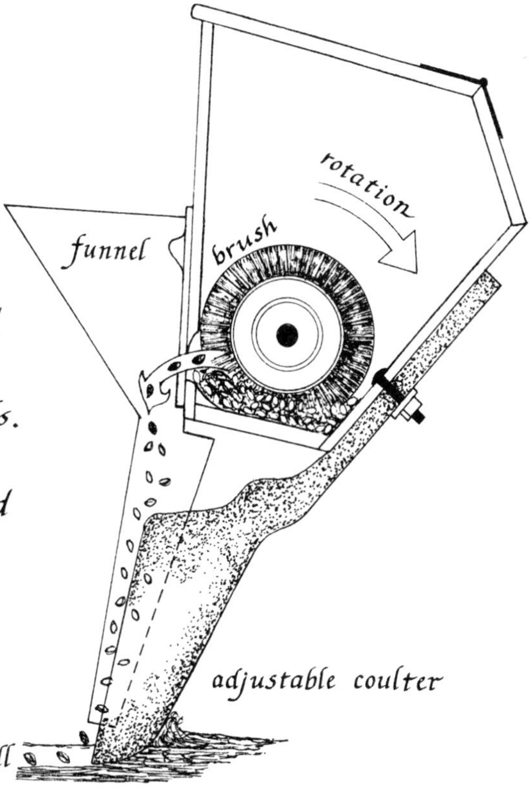

funnel

brush

rotation

adjustable coulter

drill

SECTION of BRUSH FEED BARROW

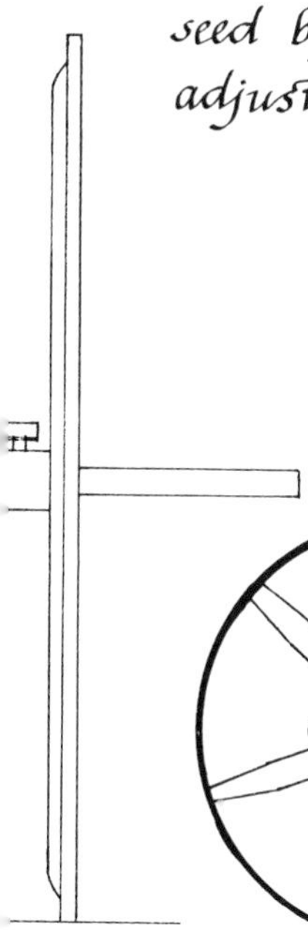

C18 Single wheel, crank driven, SEED BARROW ~ RUTLAND

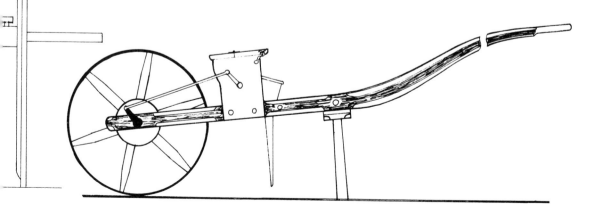

CUP-FEED SEED DRILL

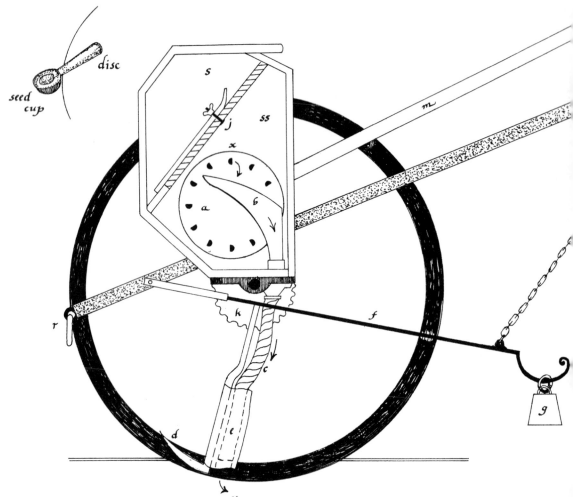

An important advance was made in 1782 when James Cooke
patented his cup feed system. This method of delivery was precise
and stood the test of time. A sectional view above of a single row
drill shows how it worked. Disc (a) is mounted on a spindle
which passes through the box. Around the edge of the disc, cups are
fixed at regular intervals. As the disc rotated it raised a spoonful
of seed from the seed compartment (ss). When a spoon reached (x)
it dropped the seeds into the

funnel (b). They passed into the flexible tube (c) & fell into the drill (furrow) made by the coulter (d) at (y). On this drill the coulter has a spur & shrouds (e) which kept the drill free from soil displaced by the coulter's motion. The depth of the coulter could be adjusted by the chain. A weight (g) at the end of the coulter arm (f) applied the force which enabled the coulter to bite the surface. The drill was pulled by a horse attached to the ring (r). It is said that this drill was once worked by a pair of dogs! This drill could be moved out of gear - a useful point when turning. The drive to the disc came from gear (k) mounted on the land

Right hand view of drill showing gears.

wheel axle. Gear (k) worked wheel (l) in a contrary direction. Both (a) & (l) are on the same spindle. To move gears 'out' lever (m) was pushed down at (q). As it was pivoted at (p) a downward movement at (mm) caused wheel (l) to rise out of gear. The speed of seed delivery from hopper (s) into the seed compartment (ss) was controlled by a slide that can be adjusted by bolt (j). Before seed drills became widely used in the C19 seeds were set by hand with dibblers, a method that employed a lot of labour.

HORSE DRILLS

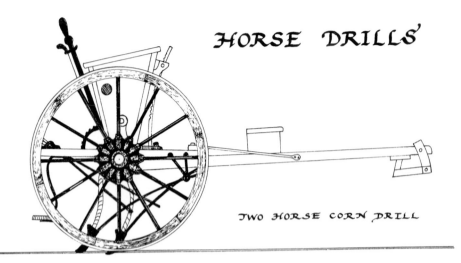

TWO HORSE CORN DRILL

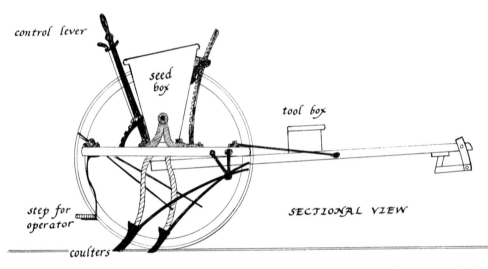

control lever

seed box

tool box

step for operator

coulters

SECTIONAL VIEW

Once a mechanical method of sowing seed had been perfected it was possible to contrive horse drawn drills capable of sowing several rows at the same time. As the C19 progressed the size of drills increased. There were many manufacturers of corn drills. In 1894 Thomas Holyoak, of Narborough, Leicestershire, sold his twelve row drill (6ft. 6 ins wide) for £27-0~0. At that time W. Rainforth & Sons, of Lincoln, produced a sixteen row drill (8ft. 6ins. wide) for £31-0~0. Testimonials appearing in advertisements show that English machines of that period were exported as far afield as Australia and New Zealand. In the post 1914 era many horse drills were converted for use with tractors.

HORSE HOES

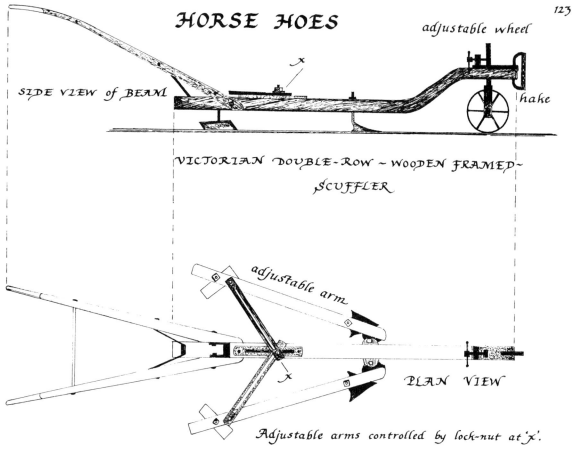

SIDE VIEW of BEAM

adjustable wheel

x

hake

VICTORIAN DOUBLE-ROW ~ WOODEN FRAMED ~
SCUFFLER

adjustable arm

x

PLAN VIEW

Adjustable arms controlled by lock-nut at '*x*'.

A regular row, sown with the precision of the machine, made
it possible for the agricultural engineer to produce a tool that
allowed the farmer to weed mechanically. This method gave
rise to great savings in terms of labour, which was always
an attraction for the prudent farmer. Various designs of
horse hoe were produced. The examples shown here were used
for roots and beans. They cleaned the ground without
burying the crop.

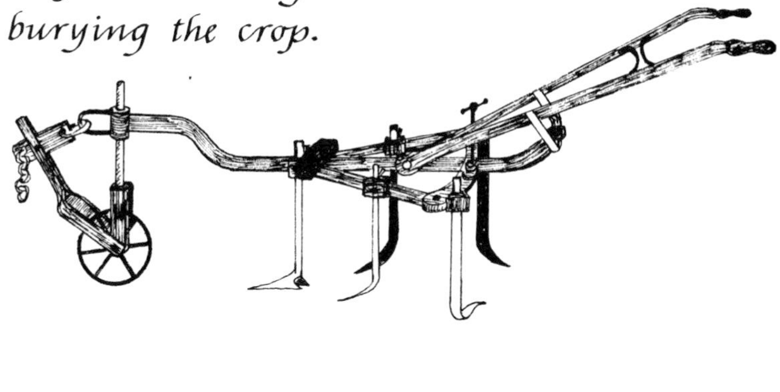

IRON ~ Single Row ~
HORSE HOE

DIBBLERS

A dibbler made a hole to admit a seed.
It was more precise a method than broadcasting. The machines which replaced dibbling
on the farmer's land did not make dibblers
obsolete. They remained useful implements in
the labourer's hand and he went on using
his dibbler on the allotment.

When corn was dibbled, and accurate dibbling
was an art, the dibbler used a pair together. Some
walked backwards and could sow a straight row
by instinct. Like many country things men often
made their own tools and these always had the mark
of natural individuality - (h & f). Corn dibblers are
not so numerous these days (h & j) outside museums.
The importance of the potato to the cottager probably
accounts for the numerical superiority of potato
dibblers - (a, b & f) are home made. Examples (c, d)
are manufactured. Specimen (e) was contrived
from an old spade handle. Turnips were sown with
(g). Almost anything long can become a dibbler
~ such as (h) which is really a game-
keeper's alarm. Its heavy weight (x)
still makes it useful for making holes
for beansticks.

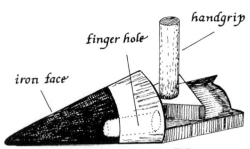

iron face · finger hole · handgrip

A CURIOUS DIBBLER

BIRD STARVING

clappers

rattle

One of the most hated and demanding tasks about the farm was that of keeping the birds away from the growing corn. It was a very lonesome occupation as one scarer was sufficient for many acres. Bird scarers were mostly children as they could be employed for a few pence a day. Even when education became compulsory, in 1870, every penny was important to labouring families. Log books kept by country schoolmasters show us that the law was often ignored and references to absentees being away bird scaring are commonplace. Lonely and cold, the child had to keep moving. Anyone found resting received a sharp reminder from the farmer's stick. Several forms of scarer were used in addition to the child's voice. The simplest was the clapper made from three pieces of wood laced together. An alternative was the iron rattle with its discs threaded on a spindle. The most effective tool was the rattle with its toothed wheels that clicked upon springy lengths of beech. Scarecrows, those almost silent sentinels of the sowing, were used even in the C18, but those shown here date from the 1930s to the present time.

rattle

HAY & HARVEST

THE SCYTHE

The scythe was not the first tool invented by man to cut the corn, but it is at least as old as the Roman era. It appears in the manuscripts of the mediaeval period and endured into this century as one of the countryman's most valued tools. In the hayfield it remained supreme until the mechanical mower came to challenge its use in the 1850s. Local custom decided its place in harvesting; where many women were employed the sickle, see page 134, was more widely used. Although machines came to replace handwork it was usually the scythe which opened up the headland for the mechanical reaper. There was always the odd corner and wildpatch where the prudent yeoman cut those extra few yards with his scythe. In some places there are still road-men who tend the verges of country roads, and where the swish of the scythe still mingles with the sounds of summer. A mower needed a sense of balance and rhythm. Those who worked all day never rushed but maintained a constant pace. A scythe was less costly than a sickle as a scytheman could cut nearly an acre of hay in a day. About the same amount of barley or oats could be cut. The sickle could reap about a quarter of an acre.

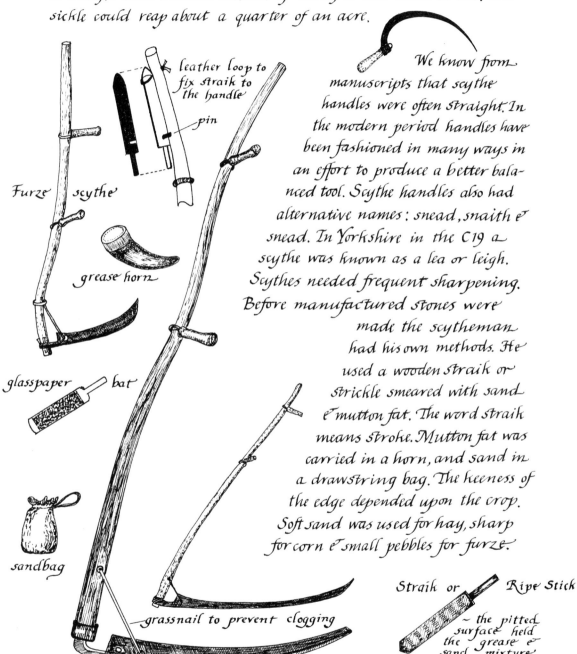

We know from manuscripts that scythe handles were often straight. In the modern period handles have been fashioned in many ways in an effort to produce a better balanced tool. Scythe handles also had alternative names: snead, snaith & snead. In Yorkshire in the C19 a scythe was known as a lea or leigh. Scythes needed frequent sharpening. Before manufactured stones were made the scytheman had his own methods. He used a wooden straik or strickle smeared with sand & mutton fat. The word straik means stroke. Mutton fat was carried in a horn, and sand in a drawstring bag. The keeness of the edge depended upon the crop. Soft sand was used for hay, sharp for corn & small pebbles for furze.

leather loop to fix straik to the handle

pin

Furze scythe

grease horn

glasspaper bat

sandbag

grassnail to prevent clogging

Straik or Ripe Stick

~ the pitted surface held the grease & sand mixture.

hauling hay with the horse

coll of threshed hay

coll of threshed hay

hayseed in sacks

riddling hayseed

barn sheet

hayseed

HAYWORK

We are apt to forget that our ancestors were prudent in their use of resources. North of the border they threshed ryegrass to extract the seed before the hay was carried home. There was a good reason for this practice. Ryegrass (Perennial) - Lolium perenne - was held to be one of the most valuable grasses from the grazier's point of view. The richness of the grazing provided by rye grass was superior to all others, and its quality was reflected in the milk yield. The drawing shows how many workers were involved in threshing. Once the hay had been cut it was tied into sheaves and set into stooks to dry. As soon as it was ready the work of threshing began. A door was used in conjunction with a fieldgate. Both were set on a cushion of hay.

DIAGRAM TO SHOW METHOD OF THRESHING HAY

haycock

raker

building a coll

flailer

forking threshed hay

forking hay to flailers

threshed hay

sheet

taking threshed hay from fieldgate

flailer

A sheet was spread beneath the gate to catch the seed. One fieldworker loosened the sheaves and pitched them upon the door. Two men stood by the door with their flails. When the flail had done its work the hay was removed and shaken over the gate. The loosened seed fell onto the sheet. The hay was gathered into a heap ready to be lifted into the new rick ~ called a coll in Scotland. Seed collected on the sheet was taken to the riddler, and after cleaning put into sacks. During the winter the seed was winnowed in the barn. An acre of ryegrass would yield about 26 bushels of seed. Even in 1897 there was a considerable variation in the bushel measure; in Gloucester it was 62 lbs., at Aberystwyth 65 lbs., at Chester 75 lbs., but in Monmouth and Abergavenny it was worth 80 lbs.

HAYSLEDGE ~ SCOTLAND

ratchet windlass

Carrying hay was a lengthy business. Haymakers always had an eye on the weather. In the north and beyond the Border, where hay had been gathered into haycocks, a sledge with a winch was used to pull them one at a time onto its flat body. The platform could be tipped so that its rear edge touched the ground. Long bodied carts were developed in the C19 and versions of the design were widely used.

HAYLIFT

Ricks were often built with the help of a horse powered hay lift. A system of pulleys was used to lift the load in the grab. When the load was in the desired position it could be dropped by a quick release device worked by the man on the ground.

LONGCART C19.

RADNOR ~ SLIDE CAR

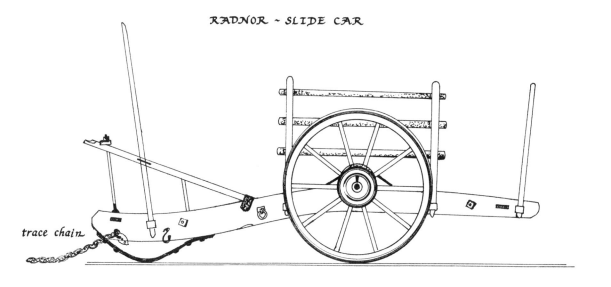

trace chain

In the hills of Wales the steep slopes prevented the use of traditional carts, and the slide car without shafts was commonly used. The horse's trace chains were fixed to the hooks at the fore end. Poles at each corner allowed a bulky load to be carried. The low centre of gravity enabled a slide car to be moved safely on steep land. Slide cars were about 12 ft. long & 4 ft. wide. Bodywork was blue and the wheels red. Slide cars were also known as 'smell posts'.

When hay was sold it was made into bales with a hay press.

HAYPRESS

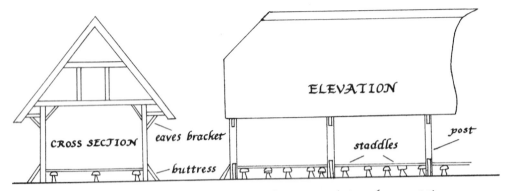

CROSS SECTION eaves bracket ELEVATION staddles post buttress

Cross section and side elevation of timber-framed hay barn with open
sides. Its removable floor stands on staddles. The roof is thatched. Extra
braces made the eaves very deep to protect the crop. Each post was made
secure by the addition of a short buttress. Buildings of this kind were
probably used at an early date, although original examples have not
survived. Steel and corrugated iron were combined in the C19 to
create the familiar Dutch barn.

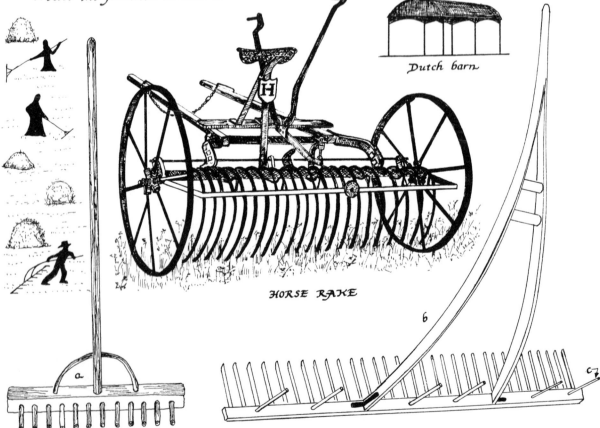

Dutch barn

HORSE RAKE

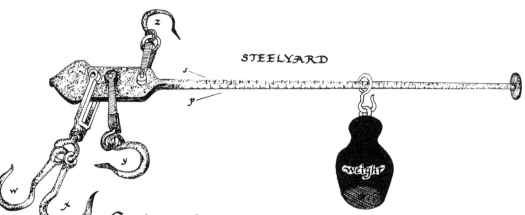

STEELYARD

Gathering hay by hand was done with a rake. Women used the small rake (a). The larger example (b) reflects the pattern in use during the early mediaeval period. Its tines are horizontal to the ground, and to stop hay spilling over the head a series of vertical pegs (c) were also attached. The head of this type of rake was about six feet long. An even larger amount of hay could be gathered by the horsedrawn hay sweep. This style of implement can be seen in most farming museums. When hay was sold it was weighed on a steelyard. The one drawn above could be used in two positions. It has four hooks: (w,x) make a pair to carry the load. Hook (z) is a suspension hook that is shown in its correct working position when scale (s) on the arm would give the weight. If the alternative scale (p) was employed suspension hook (y) would be above the arm. In many areas some hay was gathered into the barn. An unusual timber-framed and open-sided barn, shown on the opposite page, was once to be seen in Berkshire. To aid ventilation the floor was placed on staddles.

HAY SWEEP

trace chain to horse

THE SICKLE

From prehistoric times man has used a sickle to cut the corn. In the early days farmers' sickles were smaller and rounder in shape than their counterparts of modern times. The reaper in the drawing is using a large hook and a stick to hold the corn in the correct position. The use of these hooked sticks was common and in the Cotswolds they were called pick thanks. Women used sickles which were lighter tools and being smaller they cut less at a stroke. Even the advent of the reaping machine did not eliminate the use of the sickle or the reaping hook which continued to be used even into this century. Where the countryman continued to grow his own corn on his allotment it had to be harvested by hand, and this is one reason for the survival of this ancient tool. In use the sickle which had a toothed cutting edge (it was called a toothed hewk in the north) was drawn through the standing corn. This was not so with the reap hook which was used with a slashing motion. Work with the larger tool was done at a brisker pace. The shapes of the four ancient sickles show us that they were not used with a sweeping motion. The way of fixing the sickle to the handle is interesting. Its blade was extended and folded round the handle to make a clasped ferrule on the Iron Age specimen. The Bronze Age tool has a socketed head. The Roman version was flat and fitted into a slot made in the handle ~ a rivet then made it secure. A tanged end on the mediaeval sickle anticipates the usual method of fixing on the modern tools. The ancient tools were very small ~ the blade of the Bronze Age tool is about seven inches long. The task of harvesting must have been slow, and painful on the hands, in those days. Mutton fat, and some- times sand, was kept in a horn container. This mixture used on the ripe stick provided the honing agent. Corn sticks could be cut from the hedge but iron alternatives were also made. The fagging hook was also a general purpose tool & was often used for weeds.

Iron Age

Bronze Age

Roman

Mediaeval

Sickle

Bagging or Fagging hook

Strickle horn

Ripe stick or Strickle

Pick thanks

Sheaf hook

Fagging hook

HARVESTING

Harvesting was always a serious matter. For the farmer it was one of those critical times of the year when he knew that the season's fortunes were dependant upon the weather. As far as villagers were concerned the wages won from harvesting formed an important part of the family budget. Harvest money was usually boot money; & strong boots were a necessity for the man who spent most of his working hours outdoors. Itinerant Irishmen plied their services too in the harvest season, but the use of new machines gradually saw the end of casual workers. The use of the sickle seems to have lasted longer in parts of the north, where cutting corn with it was called hand shearing. To accomplish this the corn was gripped with the left hand so that the stalks could offer sufficient resistance to the hewk as it was drawn through them. The result was slower to achieve, but left a cleaner stubble. A smooth edged hook, like the one being sharpened here, was used with a 'striking' action. The left arm was then free to gather an armful of corn to set down for the bindster. About the Cheviots harvesters were divided into groups of four to six shearers. Each group was called a bandwin.

A bindster who used to bind the sheaves served two bandwins, and also sharpened the sickles for the women. The bandwin was also the breadth of corn which provided sheaves for a binder. As the shearers progressed one would make bands for the bindster and leave them on the ground. A band was made by giving a handful of corn a double twist near the head. Two shearers would each lay an armful of corn across the band which was then ready for tying. In the Border country shearers worked from west to east. The leading bandwin was at the north end of the field and its leader was the chief man. If anyone worked ahead of the leader's team the shearers could begin to race and corn was badly cut. When tempers became frayed in this way the field was said to be "on the hemp".

To catch the maximum sun stooks were set up to face N.E. & S.W. In wet weather stooks were 'hooded' with two sheaves to keep out the rain. There are some places in England, e.g. Devon, where the old style long strawed crops are still grown. The straw is used by thatchers. It is harvested with the old binding machines, which put the sickle out of business and made a great impact on rural life more than a century ago.

SCYTHE & CRADLE

A cradle gathered each swathe & left it ready for the bindster. Its use saved ten hours' labour per acre.

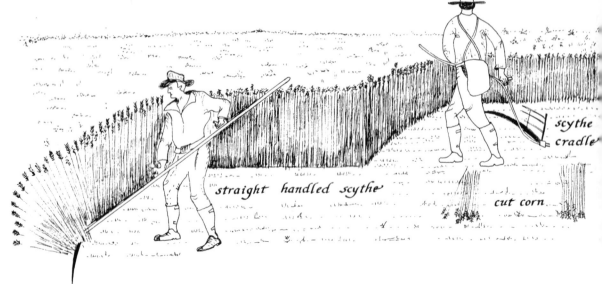

straight handled scythe

scythe cradle

cut corn

Where scythes were used to cut corn there were still many people to be seen afield, but not so many as hand shearing required. The method of working was very similar of course, but in place of several shearers scythemen worked their way across the corn. One man worked ahead of the other in the way shown above. As the corn fell in a row to the mower's left the gatherers followed to lay the bands and leave an open sheaf for the bindster to tie with one of those deft sheaf knots. The technique of tying was a matter of local usage as Tom Hennell has shown - op. cit. Nothing was left to waste in the old days and the loose straws missed by the gatherers were collected by the stubble rake. When the ground

gathering corn into open sheaves for the bindster

binding sheaves with a straw band

had been cleared the sheaves were set up in stooks to dry. In different parts of the country varying numbers of sheaves were used to form the shock, stook, stowk or hattock. Larger numbers of sheaves were placed together to make a 'mow' which gave more protection from the wet. Twenty four sheaves together made a 'thrave'

stooking

in the north. There were many customs connected with harvest. Loose ears which escaped the gatherers were once left for the gleaners. Gleaning, or leasing, is as ancient as the Old Testament, but it seems to have disappeared from the English scene a long time ago. Gleaning could not begin until the last sheaf had been carried. It was known as the 'policeman'! Various rituals attended the bearing of the final sheaf. They had definite pagan origins. To preserve the Corn Spirit the last sheaf was treated with reverence. Before being borne aloft on the last wagon it was bedecked with ribbons. This 'kirn dolly', 'mell sheaf', 'maiden' or 'neck' was kept until the following harvest, and it had a place of honour at the harvest supper. There were other names too for the maiden. It was called 'clyack' or 'cailleach', meaning the old woman, north of the border. At Duxford, Cambridgeshire, women used to throw water on the returning harvesters.

using a heel rake

wimble for twisting straw or hay bonds

REAPING MACHINES

In the early years of the C19. several engineers set their minds to the task of perfecting the mechanical reaper. The first designs cut the corn and delivered it at the rear of the machine. This was a disadvantage as the path made had to be cleared before the machine completed the journey round the field, and ran over the corn it had cut. With the development of the side-delivery reaper this problem was solved. The sails of this reaper pushed the corn to the side and clear of the machine's tracks. Reapers of this kind could cut about an acre an hour working round about the standing crop. Some six to ten workers were needed to follow the reaper and to bind, lift and stook.

BINDERS

The next stage in the mechanization of the harvest was the development of the binder. This cut the corn by pushing it downwards with its revolving sails (s). Each cut fell onto the platform (a) and the moving surface carried it towards the belt (b). It was then carried upwards, between the belts x & y, to (c) where it passed onto the downward slope (d) to be collected into a sheaf and bound with twine. It was then discharged at (e) for stookers to collect.

The main advantage of the binder was its speed of working. This made it possible to harvest crops much more quickly than the old hand methods allowed. Victorian farmers were also aware of the hazards of bad weather and recognised the real advantages of quicker methods. In recent years the demand for long straw for thatching has seen the revival of the binder. Binders can once again be seen with their sails slowly turning behind the tractor, which now replaces the horse as the machine's motive power. The binder appeared in the 1880s and was a feature of the farming scene for at least forty years until it was challenged by the combine harvester in the 1920s.

Stooking was always thirsty work and those who went afield did not forget to take some refreshment with them. Cold tea or cider in a costrel were the traditional beverages. Both were administered with a horn beaker or enamel mug.

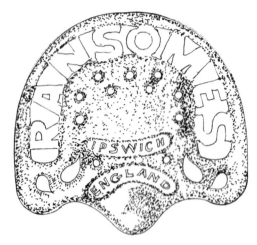

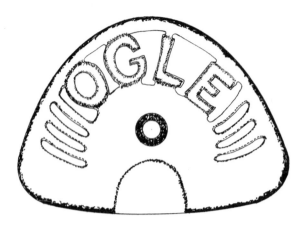

MACHINE SEATS

After the introduction of the horse drawn machines, like rakes & reapers, manufacturers soon took the opportunity to make their products advertise the company name. Machine seats became advertisements. Many seats seem to have survived although the implements have disappeared. Victorian designers showed great inventiveness and seats were made in various shapes. Cast iron & pressed steel was used. Care was taken with the details of lettering. The style used by Albion is perhaps one of the most elegant as the drawing shows. This durable form of the advertisers' art is now a very collectable item.

ALBION

RICKING

corn rick set on staddles

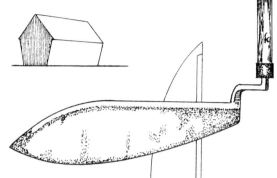

Two kinds of ricks were built. Corn had to be protected from rodents by standing on stone or latterly iron staddles. Hay stood on the ground. A corn rick was made by laying branches across the staddles and placing the sheaves on top. Staddle frames of iron were also made for this purpose. Long before iron pitchforks (a) were mass produced countrymen devised their own (b) from the hedgerow. It was important for the farmer to know the quality of the hay or corn inside the rick, and sectional rods (d) were made for sampling the interior. Different ends (e, f & g) were made for hay or corn. The number of intermediate rods (h) used depended upon the size of rick. Hay becomes compressed & when a rick was opened up large hay knives were needed to slice through the succulence the cattle were waiting to enjoy. Some knives had straight handles (j) & others had off-set ones (k) which made it easier to cut a vertical face.

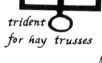

trident for hay trusses

Cornwall

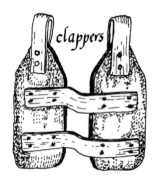

clappers

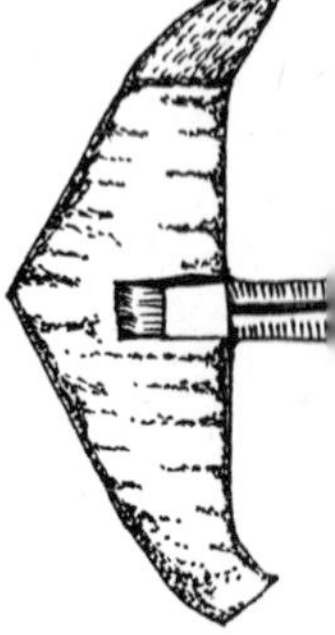

BREAST PLOUGHS

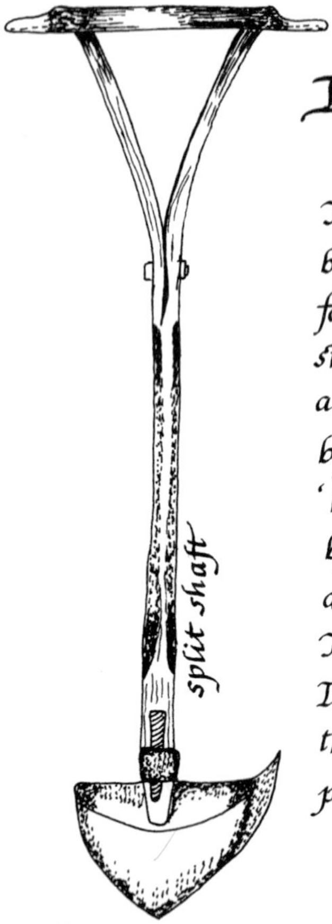

The 'most slavish' task in husbandry was burnbaking. Before the days of pesticides farming methods required that the pared stubble was burnt. Paring was done with a breast plough which the operator pushed before him. To protect the thighs 'clappers' or 'beaters' were worn suspended from the belt by two looped straps. A man could clear about half an acre a day with such a tool. The beam of the plough is about six feet long. Its upturned edge, called the counter, served the same purpose as the coulter on a horse plough. In use the point, the picket, often needed sharpening.

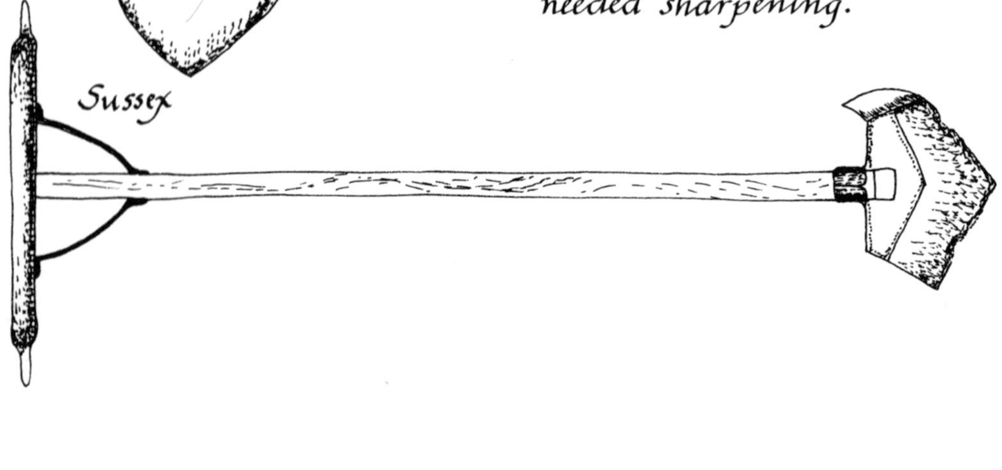

split shaft

Sussex

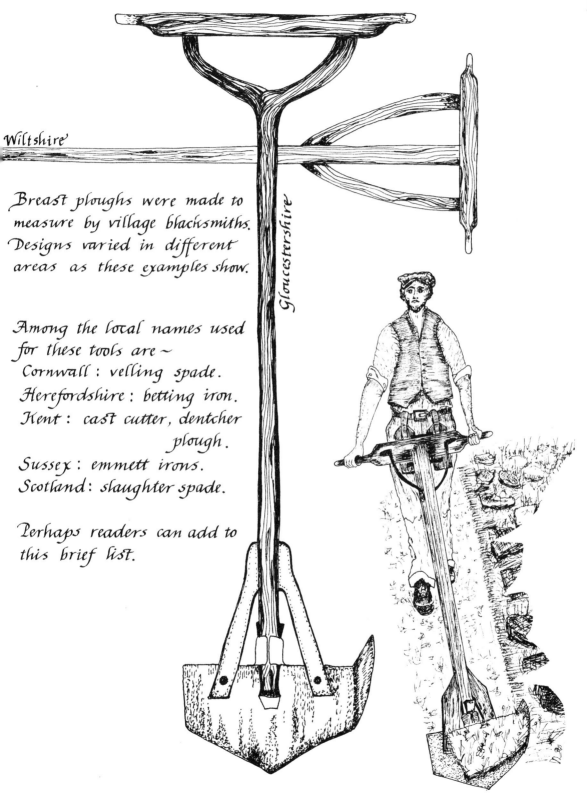

Wiltshire

Gloucestershire

Breast ploughs were made to measure by village blacksmiths. Designs varied in different areas as these examples show.

Among the local names used for these tools are ~
Cornwall : velling spade.
Herefordshire : betting iron.
Kent : cast cutter, dentcher plough.
Sussex : emmett irons.
Scotland : slaughter spade.

Perhaps readers can add to this brief list.

SHEEP, SMOCKS & TRAPS

A shepherd's work was never really finished. Sheep need constant care, & there are many tools which the shepherd once used in his work that, today, are outdated. Antique dealers sell them at prices that would have amazed their original owners. Bells were fixed with collars or with a yoke. Bells made from iron sheets were sometimes dipped in brass to improve their tone. Old bells are often mottled where brass has worn away. Sheep washing came a few days before the annual shearing. Dip hooks with a cyma curve were common. Shears were often handed down as heirlooms. The example shown bears the date 1779 on the other side of the sheath. Ram yokes were used to prevent fence breaking. The tar pot & marking irons were essential when sheep roamed free. Lambs' tails were docked with an iron, & the wound cauterised with a searing iron.

yoke & bell

shears

·I·S·
1844

leg crooks

ram yoke

dip hooks

rubbing bar

COOPER'S DIP

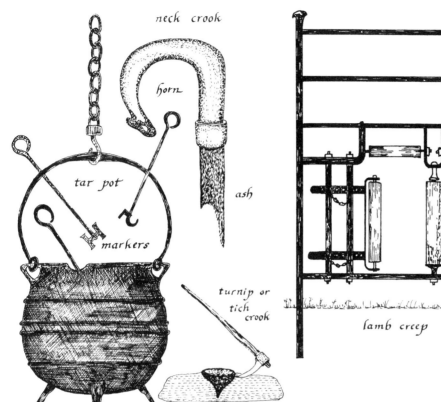

neck crook

horn

ash

tar pot

markers

turnip or tich crook

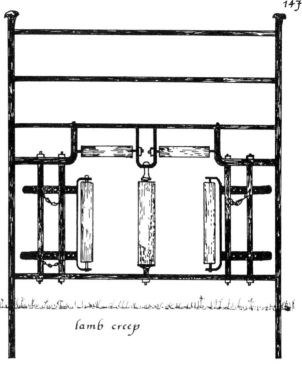

lamb creep

drench horn ~ used to administer salves

docking iron

searing iron

The shepherd's principal symbol was the crook. There were two types of crook. Those with a small iron hook were made to hold a sheep's leg. The larger example above is made of horn, and can be used to hold the sheep's neck. In the middle ages the woolmerchants too made use of the crook emblem. The sheep below stands on a woolsack marked with two leg crooks. It appears on a C15 brass at Northleach, Glos.

Another useful tool was the tich crook used for hooking the uneaten portion of a turnip out of the ground after the flock had grazed the upper part. A lamb creep allowed them to get a first bite on new grazing. It was set in a line of hurdles.

The task of moving the flock from one grazing area to another on a field of roots took a lot of time. Hurdles were used as temporary fences. Their pointed legs enabled them to be driven into the ground. A concave-headed fold beetle could be used for this purpose. An alternative method was to use a fold bar which also had a concave part that could fit over the end of a hurdle stake (a). To drive a leg home a large wooden beetle was employed. The advantage of a fold bar was that a hole could be made to admit the end of the stake with its point.

Shepherds needed to keep warm. This design of firebox was common. Its curved horns allowed it to be suspended upon a horizontal iron bar.

Treating sheep out of doors was made easier by using a yoke that could be pushed into the ground. The pin was adjustable. To protect the sheep the inside of the yoke was leather lined.

dip hooks

firebox

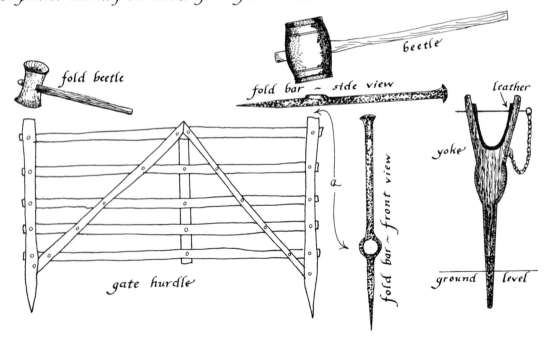

fold beetle

beetle

fold bar ~ side view

leather

a

fold bar ~ front view

yoke

gate hurdle

ground level

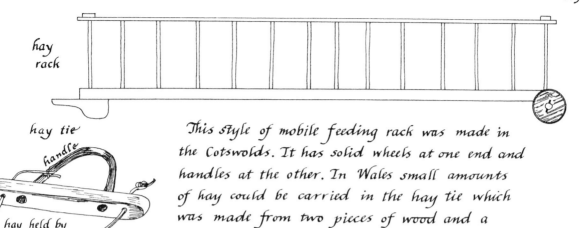

hay rack

hay tie

handle

hay held by string

This style of mobile feeding rack was made in the Cotswolds. It has solid wheels at one end and handles at the other. In Wales small amounts of hay could be carried in the hay tie which was made from two pieces of wood and a stout piece of string to gather the hay.

Old style shepherding needed a constant supply of hurdles. The end stakes of a woven hurdle were called shores, and the intermediate members sails. A hole at the centre allowed several hurdles to be carried together on a pole.

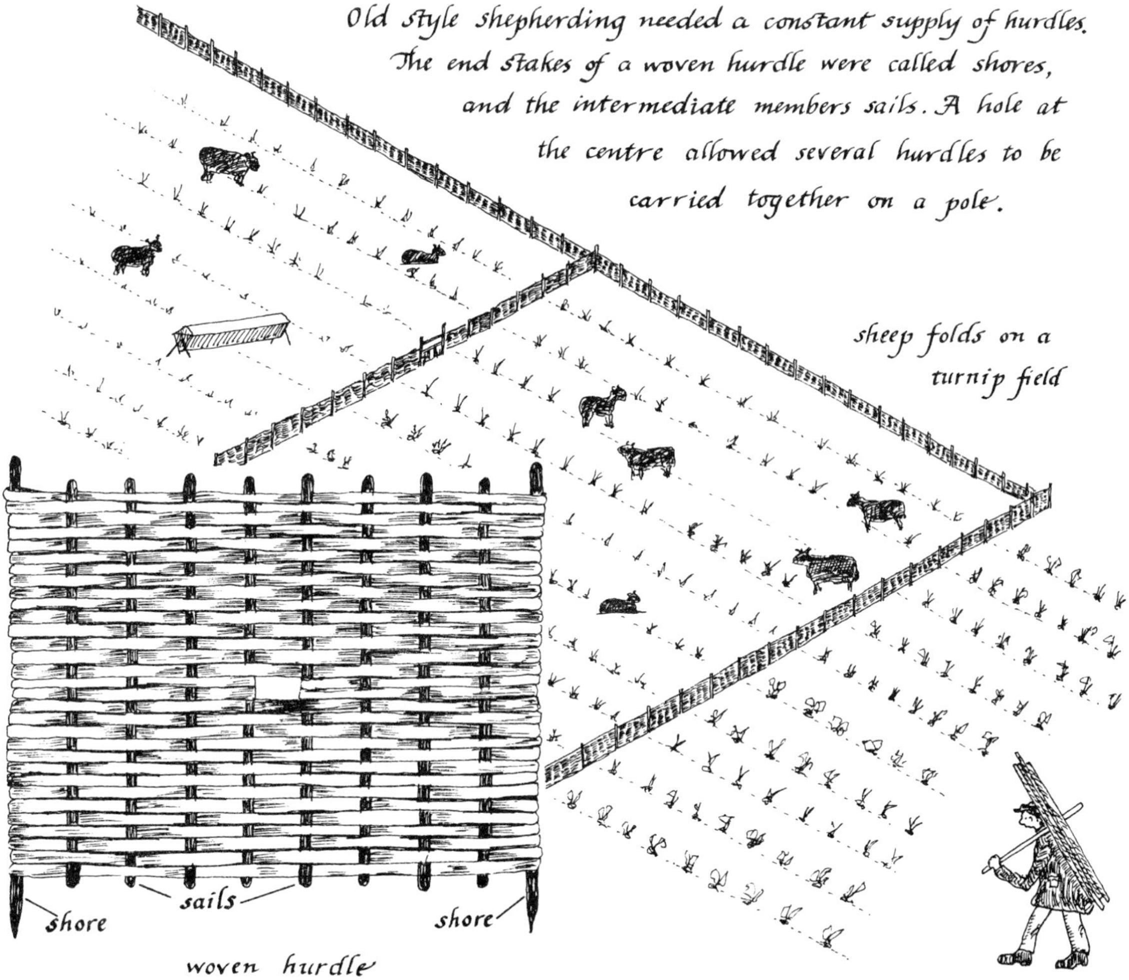

sheep folds on a turnip field

shore   sails   shore

woven hurdle

The portable shepherd's hut was once a familiar sight. Each lambing season it was taken to the downs to be a shelter & a dispensary for the shepherd. This hut has iron wheels. Its under-carriage has a pole & braces like a wagon. The steps hook into staples below the doorway. To give an all round view of the flock there are windows on three sides. A stable door could be left half open so that the shepherd could hear as well as see his flock even when he was resting within.

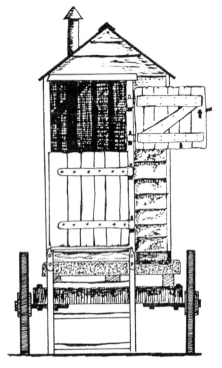

MOLE TRAPS

Loops inside the mole trap

string to spring

spring

BARREL MOLE TRAP in the set position

3

BARREL MOLE TRAP in released position

3

Moles were a nuisance to some and a source of income for others. There are various kinds of trap. The simplest type was a wooden run 1 which had a trapdoor that deposited the mole in a jar of water. The spring trap 2 with its long spikes killed the mole ~ Talpa europaea ~ but the holes made in the skin made it unsaleable.

Barrel traps 3 were effective but not much fun for the victim. The spring was set by placing a wedge (w) in the hole (h). When the mole passed through the barrel the wedge was dislodged and the string (s) jerked upwards. Two loops (l) caught the mole as the drawing shows. Barrel traps were usually made of wood, but example 4 was made of baked clay. The initials are those of Thomas Turner who lived at Tingewick, Buckinghamshire. He obtained 6d a skin for his moles about a century ago.

2

Mole spade used for setting traps

Paddle~shaped mole spade.

handle about 4ft long

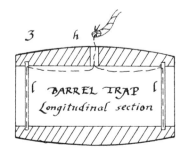

3   h

BARREL TRAP
Longitudinal section

4

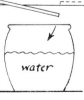

1

water

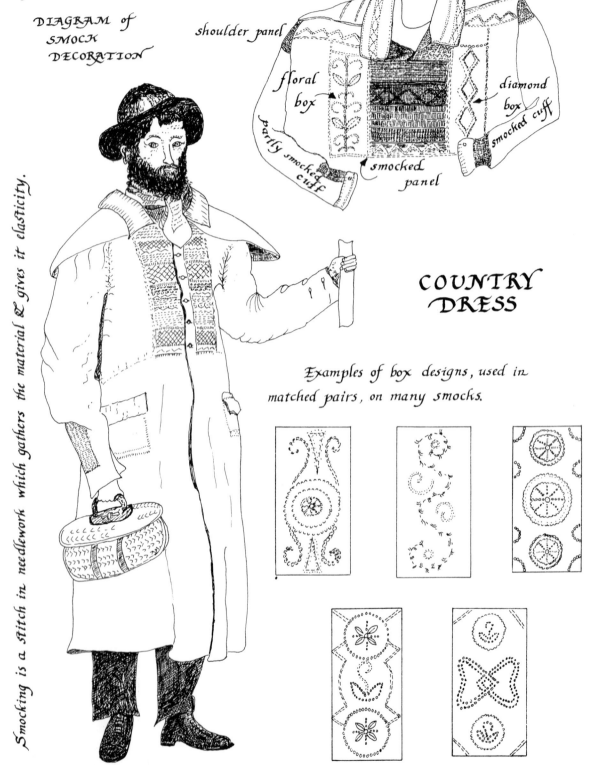

DIAGRAM of
SMOCK
DECORATION

shoulder panel

floral
box

diamond
box

partly smocked
cuff

smocked cuff

smocked
panel

COUNTRY
DRESS

Examples of box designs, used in
matched pairs, on many smocks.

Smocking is a stitch in needlework which gathers the material & gives it elasticity.

To keep off heavy rain those afield often wore a sack, slit down one side, over the head. It made a hood & saved the back from driving rain. The characteristic county garment, however, is the smock. They were originally overalls to guard the wearer from the elements and clothes from harm. Smocks were used by menfolk and children. Originating in the late C18 they were in common use for about a century, and then fashions changed. In the West Midlands & Wales smocks had large collars which gave extra protection from the wet. Boxes on each side of the front smocking displayed a variety of decorative emblems. It is doubtful if this form of embellishment indicated the wearer's occupation. Smocks varied in colour. Sunday & wedding smocks were white, but working colours included unbleached materials & grey, brown, green & blue. In William Howitt's 'Boys' Country Book' the boy scaring birds wore a blue smock – perhaps he was the inspiration for Little Boy Blue. Thomas Hardy – 'Under the Greenwood Tree' – also mentions the hearts, diamonds and the zig-zags on the carol singers' smocks. Material to make a smock cost 4½d. a century ago and the smock maker was paid about 9d. for all the work involved. The usual linen fabric used was called Holland; and in its unbleached form was known as brown Holland.

Women fieldworkers wore long dresses and aprons. Their heads were protected by bonnets with deep collars to keep off the sun & rain.

side view   iron

PATTENS were worn over shoes to keep the wearer's feet clear of standing water.

THE LABOURER'S HARVEST

In the C19 field labour was easy to come by as far as the farmer was concerned. Once the relative security of stripholding had been taken away by enclosures ordinary folk had to depend on day-work. Wet weather meant no pay. 'Estate' villages with a caring owner provided more regular employment, but most countrymen had to rely upon their own enterprise to maintain their families. Allotments therefore played an important part in the domestic economy of the farmworker. They had to be dug by hand ~ sometimes by lantern light. The tools discarded by the farmer, which had served him before the age of the machine, found their way to the labourer's allotment, and remained in use well into the present century.

Harvesting had to be fitted in around the demands of the employing farmers. In some villages when all the allotment holders had gathered their corn it was put into a communal rick. Each man's harvest was separated by a layer of straw and a marker defined ownership. A contractor's machine was hired to do the threshing. After each layer had been fed into the box it was left to run dry so that one man's corn was threshed before another's was begun.

1. Weeding tongs. 2,3,4. Dock lifters. 5,6. Paddles for cleaning a plough & 'spudding' weeds. 7. Hoe. 8. Draw hoe ~ with a cutting edge inside the hook.

POSTSCRIPT

In a changing world there are still some old farms which can claim a family ownership, or tenancy, of more than a century. Most old style farmers seldom discarded anything ~ never knowing when it might come in handy. To their prudent practices we owe much of the artefact material which has survived to our own time. There are still treasures to be found for those with the determination to search them out. This book has described a selection of bygones from farmstead and field. The author will be pleased to hear of other examples worthy of attention.

156

PLACES TO VISIT

Some of these places are only open during the summer. Readers are advised to check specific opening times.

Abergavenny Museum, The Castle, Abergavenny, Gwent.

Avoncroft Museum of Buildings, Stoke Heath, Bromsgrove, Worcs.

Buckinghamshire County Museum, Church St., Aylesbury, Bucks.

Butser Ancient Farm Project, Queen Elizabeth Country Park, Gravel Hill, Horndean, Hants.

Cotswold Collection, The Old Gaol, Northleach, Glos.

James Countryside Museum, Bicton Gardens, East Budleigh, Devon.

Kirbee Rural Crafts Museum, Whitchurch, Ross-on-Wye, Hereford. (Visits by appointment)

Manor Farm Museum, Cogges, Witney, Oxon.

Museum of East Anglian Life, Stowmarket, Suffolk.

Museum of English Rural Life, University of Reading, Whitenights, Reading, Berks.

Naseby Farm Museum, Purlieu Farm, Naseby, Northants.

Norton's Farm Museum, Sedlescombe, Battle, Sussex.

Old Kiln Museum, Reeds Rd., Tilford, Farnham, Surrey.

Rutland County Museum, Catmos St., Oakham, Rutland.

Smerrill Farm Museum, Kemble, Cirencester, Glos.

Sutton Windmill, Stalham, Norfolk.

Weald & Downland Open Air Museum, Singleton, Sussex.

Welsh Folk Museum, St. Fagans, Cardiff, S. Glamorgan.

Wiltshire Rural Life Centre, The Great Barn, Avebury, Wilts.

ACKNOWLEDGEMENTS ~ The author is pleased to record his appreciation for the help he has received from the above collections & from the following: Thora Ansty, Henry North, Catriona Nicholson, A.J. Kirby, R. Merson, Leslie Richards, Gerald Anderson, Mike Griffin, Eddie Lambourne, Henry Jackson, Owen J.M. Lee, Ronald Stiff, Dr. Sadie Ward, Ann Williams, Chris. Walker, Pamela Clement, Christopher N. Gowing & Roger Hudson.

INDEX

Delta (standing),
Dodger and Lisa Marie